AF589386

Fish Products and Value Addition

THE AUTHORS

Dr S Balasundari is working at Tamil Nadu Dr. M.G.R. Fisheries College and Research Institute, Tamil Nadu Dr. J. Jayalalithaa Fisheries University at Ponneri as Professor and Head, Department of Fish Processing Technology, having 25 years of professional experience that includes her service in fish processing industries, ARS and SAU. She is a member of AFSIB, Agricultural Scientific Tamil Society, SOFT and WAS. She served as a production executive in Seafood Processing and Export Companies for three years, possessing sound knowledge on processing and export of shrimp, crab, lobster, squid, cuttlefish and finfish. She has developed various contemporary value added fish products and disseminated the technologies. She has completed six projects through state and national level funding agencies and established demonstration units for the benefit of stakeholders.

G Raghu is working as Teaching Assistant of Tamil Nadu Dr.J.Jayalalithaa Fisheries University, Nagapattinam. He has 3 years of teaching experience in various aspects of fish quality and fish processing and 2 year field experience in aquaculture sector. He has completed his Master of Fisheries Science in Fish Quality Assurance and Management with TNFU Merit Fellowship at Fisheries College and Research Institute, Tamil Nadu Dr. J. Jayalalithaa Fisheries University, Thoothukudi. He is the member of society of Fisheries Technologist and Professional Fisheries Graduates Forum in India. In addition, he has participated in National and International Conferences, Symposia, Workshops, Seminars, etc., organized by fisheries development organizations. He has developed one technology on "Antioxidant edible film from fish gelatin and fucoidan". He has filed one patent entitled "Method of Antioxidant releasing edible film from carp gelatin and fucoidan" (Application Number: 201641031823). He has published research and extension articles in national and international journal.

Dr S Felix, Ph.D., Vice Chancellor, Tamil Nadu Dr. J. Jayalalithaa Fisheries University, Nagapattinam is having experience of more than 35 years in the area of fisheries and aquaculture in teaching, research, extension and administration. He has organized more than 20 Nos. of seminar/conference/symposium at National and International levels. Currently he is also the President (2018-19) of World Aquaculture Society, Asian Pacific Chapter and Chairman, ICAR's BSMA Committee (2018-19). He has more than 60 research papers published in National and International journals. He has authored more than 10 books and 25 manuals. He is specialized in the area of advanced systems in aquaculture and aquariculture such as raceway, biofloc technology, RAS, etc,. He has operated around 20 externally funded research projects.

Fish Products and Value Addition

by

S Balasundari

G Raghu

S Felix

Tamil Nadu Dr. M.G.R. Fisheries College & Research Institute
Tamil Nadu Dr. J. Jayalalithaa Fisheries University
Ponneri, Tiruvallur District, Tamil Nadu

DAYA PUBLISHING HOUSE®
A Division of
ASTRAL INTERNATIONAL PVT. LTD.
New Delhi – 110 002

ISBN: 9789388173247 (PB)

Publisher's Note:

Every possible effort has been made to ensure that the information contained in this book is accurate at the time of going to press, and the publisher and author cannot accept responsibility for any errors or omissions, however caused. No responsibility for loss or damage occasioned to any person acting, or refraining from action, as a result of the material in this publication can be accepted by the editor, the publisher or the author. The Publisher is not associated with any product or vendor mentioned in the book. The contents of this work are intended to further general scientific research, understanding and discussion only. Readers should consult with a specialist where appropriate.

Every effort has been made to trace the owners of copyright material used in this book, if any. The author and the publisher will be grateful for any omission brought to their notice for acknowledgement in the future editions of the book.

Published by	:	**Daya Publishing House®** *A Division of* **Astral International Pvt. Ltd.** – ISO 9001:2015 Certified Company – 4736/23, Ansari Road, Darya Ganj New Delhi-110 002 Ph. 011-43549197, 23278134 E-mail: info@astralint.com Website: www.astralint.com
Laser Typesetting	:	**Classic Computer Services,** Delhi - 110 035
Printed at	:	**Neelam Graphics, Delhi - 110007**

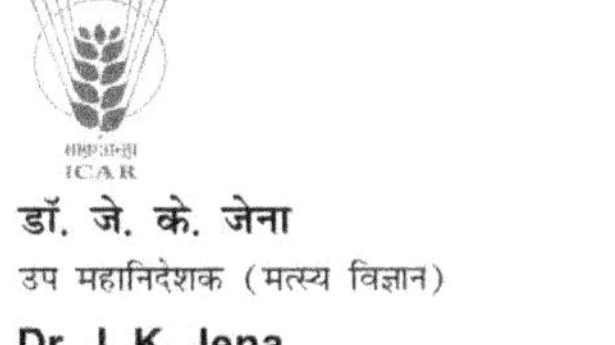

डॉ. जे. के. जेना
उप महानिदेशक (मत्स्य विज्ञान)

Dr. J. K. Jena
Deputy Director General (Fisheries Science)

भारतीय कृषि अनुसंधान परिषद
कृषि अनुसंधान भवन-II, पूसा, नई दिल्ली 110 012
INDIAN COUNCIL OF AGRICULTURAL RESEARCH
KRISHI ANUSANDHAN BHAVAN-II, PUSA, NEW DELHI - 110 012

Ph. : 91-11-25846738 (O), Fax : 91-11-25841955
E-mail: ddgfs.icar@gov.in

Foreword

The demand for fish, as an important components of the diet, is growing significantly all over the globe, due to its high nutritive value and palatability. Several processing technologies have been developed for different fish & shellfish verities over the years ensuring increasing storage duration, distance transportation, assuring quality & hygiene for the ultimate consumers, effective utilization of low-valued fish and providing higher economic benefit. A number of methods are used to preserve fish, which primarily include employing techniques based on temperature control, using ice, refrigeration or freezing; and others on the control of water activity, including drying, salting, smoking and freeze-drying. Further, the technologies of development of new and innovative value added products have added new dimensions for use of fish depending on the requirements of the diversified consumers' base. These products range from live fish and shellfish to ready-to-eat convenience products. Considering the fact that the industry is becoming highly competitive, value addition offers alternative approach for raising profitability from the enterprise.

This book entitled ***'Fish Products and Value Addition'*** authored by Drs S. Balasundari, G. Raghu and S. Felix is well written and gives significant information about principles of fish preservation and processing, traditional methods of fish preservation and diversified products from fish. I appreciate the effort of the authors in bringing out this useful publication.

(J.K. Jena)

Message

Fish is a wonderful, nutritious food that has much health beneficial attributes which helps us to overcome lifestyle diseases and malnutrition. Fish needs to be preserved and processed effectively in order to supply fish to the consumers without nutrient loss. This book deals with the subject of fish preservation techniques which is useful for students, fishers and entrepreneurs. There are entrepreneurs interested in taking up the business of production of value added fish products. Latest technologies in value addition is explained in this book to support entrepreneurship.

S. Felix
Vice Chancellor
Tamil Nadu Dr. J. Jayalalithaa Fisheries University

Preface

The post harvest losses recorded is 12% which leads to intense revenue loss to fishers. Creation of awareness among fishers, train them on improvised fish preservation techniques and value addition will be the ideal solution to overcome this loss. The reduction in postharvest losses has a direct impact on local and regional trade flows. Per capita consumption of fish is very low, it is only 9 kg against the recommendation 14 kg. Availability of diversified quality fish products in the domestic market has to be increased to ensure increase in per capita consumption of fish. The content of this book is planned to overcome these issues by providing detailed information on fish preservation methods and value addition.

Authors

Contents

Chapter 1

Principles of Fish Preservation and Processing

1.1. Composition of Fish

Composition refers to elements involved in composing. Fish and shellfish are made up of various chemical constituents.Chemical compositions of fish and shell fish makes them distinct from other animal meat.

Proximate Composition

Major Components	*Minor Components*
☆ Moisture	☆ Vitamin
☆ Protein	☆ Non protein nitrogenous substances such as free amino acids, urea predominantly in cartilaginous fishes, trimethylamine oxide mainly in marine fish and crustaceans
☆ Carbohydrate (Very less compared to terrestrial animal)	
☆ Fat (Lipids)	☆ Hydrocarbons (*e.g.* Squaline) and many other substances.
☆ Minerals (Ash)	

1.2. Proximate Composition

Proximate composition refers to relationship of the major chemical constituents in terms of proportion to the whole weight expressed in percentage.

1.2.1. Parameters Affecting the Proximate Composition of Fish

☆ Body part, species, age, sex, size, sexual maturity, habitat, fishing season.

Table 1.1: Average Proximate Composition of Fish

Proximate Composition	*Percentage (w/w)*
Moisture	60-80 per cent
Crude protein	12-22 per cent
Total lipids	0.2-2.2 per cent
Total ash	0.5-1.5 per cent

- ✰ Protein content shows less difference in different fishes, whereas lipid content shows wide variations.
- ✰ Fat content decreases from tail towards head and from ventral towards dorsal portion.
- ✰ Fin fishes differ from terrestrial animals by the presence of dark muscle. Dark muscle contains more fat than white meat in some fatty fishes

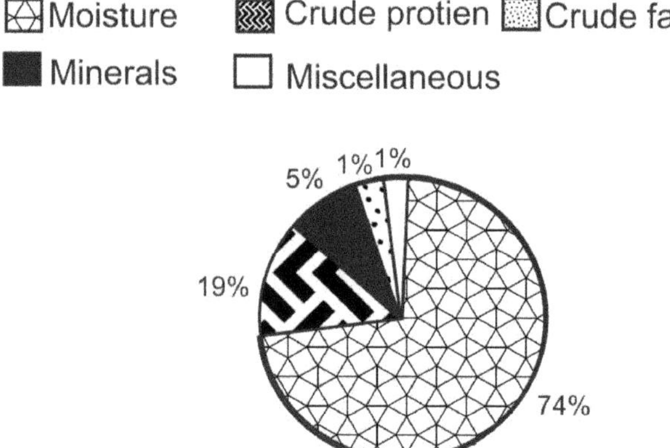

Figure 1.1. Average Proximate Composition of Fish.

1.3. Spoilage of Seafood

Physical or biochemical deterioration of breakdown of tissue makes the fish unfit for human consumption. All the food commodities are perishable. But fish is highly perishable. Because of

- ✰ high moisture
- ✰ low glycogen reservoir leads to post mortem pH near neutral favors the Microbial growth
- ✰ low connective tissue makes the major protein more susceptible to proteolytic degradation

- ✰ highly unsaturated fatty acids more prone to oxidation
- ✰ high content of non-protein nitrogenous compounds

Table 1.2: Causes of Seafood Spoilage and Intrinsic Characteristics of Fish

Causes of Spoilage	*Intrinsic Characteristics*
Bacteria	✰ High moisture
	✰ Near neutral post mortem pH
	✰ High contents of non protein nitrogenous compounds
Enzymes	Low connective tissue
Oxidation	Highly unsaturated fatty acids more prone to oxidation

Other animals' meat have post mortem pH in acid range, compared to fish muscle and high connective tissue content delay the degradation of protein by endogenous proteases as well as secreted from bacterial origin.

Table 1.3: Result of Seafood Spoilage

Spoilage Consequences	*Causes*
Reduced Shelf life	Due to the action of bacteria, enzymes and oxidation
Off flavours and smells	From the breakdown of tissue through action of bacteria and enzymes
Taints	Off flavours that arise from contamination during handling and preparation
Reduced quality	Deterioration of the visual, physical and chemical characteristics of seafood from the action of bacterial and enzymes
Food poisoning	Predominantly from the contamination and growth of bacteria

1.4. Fish Preservation Methods and Principles

At present different methods are used to preserve the fish and fishery products based on the desirable end product properties. Most commonly used fish preservation methods are; chilling, freezing, curing (drying, salting and smoking), canning, marinating, boiling and fermentation. The other methods such as preservation by irradiation, freeze-drying, modified atmospheric packaging and retort pouch packaging are also used for preserving fish.

1.4.1. Chilling

Preservation by chilling is mainly due to the lowering of the temperature of the fish as low as possible (near to 0°C) to delay both biochemical and microbiological processes. With the lowering temperature the lag period of microorganisms will be extended, resulting in delayed growth. The lower the fish temperature, the low will be the activity of enzymes and microorganisms.

1.4.2. Freezing

Fresh fish flesh normally contain 60-80 per cent water depending on the species and freezing process converts most of this water into ice. In freezing along with the effect of low temperature, mechanical rupture of bacterial cells during ice formation, freezing out of the major fraction of water in substrate also contributes to the death of microorganisms, thus by extending shelf life. The lower the fish temperature, the low will be the activity of enzymes and microorganisms. The oxidative rancidity problem is controlled by glazing the product before freezing.

1.4.3. MAP (Modified Atmospheric Packaging)

The preservation in MAP is by retarding the growth of microorganisms by changing the gaseous composition of the environment, thus by creating unfavorable conditions for microbial growth (mainly due to the effect of carbon dioxide on microorganisms) and by avoiding the lipid oxidation.

1.4.4. Curing (Drying, Salting and Smoking)

Unlike canning (which engenders the destruction of micro-organisms and their spores) curing preserves by rendering the medium unsuitable environment for microbial propagation. Increasing the concentration of soluble substances in the medium either by abstracting water or by causing soluble substances to diffuse in (salting, brining or sugar curing) are the principal means of accomplishing this. In addition to concentrating the soluble substances by brining and dehydration, smoking preserves by depositing bacteriostatic chemicals like formaldehyde and phenols in the system. The addition of salt is more effective for weight than the addition of sugar because salt ionizes to a sodium cation and a chloride anion each of which attracts a sheath of water molecules. These ionically associated water molecules are unavailable for use by micro-organisms and there is a tendency for the ionic forces to pull water molecules from the microbial cells dehydrating them to the point where they die or sporulate and lie dormant. Sucrose also withdraws water molecules from the system and holds them by hydrogen bonding. However, far fewer molecules become bound or unavailable in this way than is the case for an equal mass of sodium chloride. This availability of water in the system for use by microorganisms directly relates to the effectiveness of preservation and can be represented physically by the water activity (a_{w}).

1.4.5. Canning and Retort Pouch Packaging

The preservative effect in both the cases is mainly by subjecting the products in hermetically sealed containers/pouches, to high temperatures in order to bring the commercial sterility, where most of the microorganisms cannot survive, except highly heat stable spore formers. In this method the products are heated to high temperatures (121°C) for certain time with the intention of achieving commercial sterility to avoid the risk of pathogens and toxins, mainly *Clostridium botulinum*, which is a high heat resistant spore forming and toxin producing bacteria occurs in canned foods.

1.4.6. Marinating

The marinades are preserved by keeping them in acid medium (acetic acid and propionic acid) containing salt at a pH 4.5. At a pH 4.5 or below most of the spoilage causing and all food poisoning bacterial growth is arrested, resulting in a product with a characteristic flavor and an extended but limited shelf life. Acetic acid controls the pH and selectively allows the autolytic reactions to take place. The salt (sodium chloride) causes the removal of water and coagulates the protein. It also controls the hydrolytic action and allows it to proceed within desired limits. However some bacteria and enzymes will remain active and cause spoilage, which can be slow down by storing at low temperatures (below 10°C). The amounts of acid and salt required can be reduced when the product is kept chilled until eaten.

1.4.7. Boiling

The action of boiling fish in water at normal temperatures and pressures denatures (cooks) the proteins and enzymes and kills many of the bacteria present on the fish. The normal spoilage that occurs in a dead fish is thus stopped or drastically reduced. Often salt is added before, during or after processing; high levels of salt in the final product will help to extend the shelf life.

1.4.8. Fermentation

The fermentation processes are those in which organic catalysts (enzymes or ferments) break down complex organic molecules to simpler ones. Many of the processes used in fish preservation aim at keeping the fish flesh as near as possible to its original condition. With fermentation, however, we are considering methods by which the wet protein is broken down to simpler substances which are themselves stable at normal temperatures. In some of the processes we shall be considering, breakdown is only partial and is controlled by the addition of salt; thus the process is designed to produce a particular flavour as well as to preserve the product.

1.4.9. Irradiation

Food irradiation is the process for the treatment of food products to enhance their shelf life and to improve microbial safety. Electromagnetic radiations, namely gamma, and X-rays having short wavelength (< 300 μm) and higher energy than visible light can significantly penetrate the material including foods causing ionization of atoms and molecules by removing electrons from their outer shell. The inactivation of living cells (microbial cells) by irradiation is essentially due to scission of single or double strands of DNA, which is essentially caused by the OH radical formed by radiolysis of water. In addition to DNA damage, ionizing radiation has also been shown to cause damage to the membrane and other structures causing sub-lethal injury.

1.4.10. Freeze-Drying

Freeze drying is a dehydration process typically used to preserve a perishable material or make the material more convenient for transport. Freeze-drying works by freezing the material and then reducing the surrounding pressure to allow the frozen water in the material to sublimate directly from the solid phase to the gas phase (*i.e.*, it does not transit through the liquid state) under vacuum. Freeze-drying benefits heat-sensitive products by dehydrating in the frozen state without intermediate thaw. Freeze-drying of meat yields a product of excellent stability, which on rehydration closely resembles fresh meat. Adequate control of processing conditions contributes to satisfactory rehydration, with substantial retention of nutrient, colour, flavour, and texture characteristics.

1.4.11. Hurdle Technology

Hurdle technology (also called combined methods, combined processes, combination preservation, combination techniques or barrier technology) advocates the deliberate combination of existing and novel preservation techniques in order to establish a series of preservative factors (hurdles) that any microorganisms should not be able to overcome.

Chapter 2

Traditional Methods of Fish Preservation

Traditional Methods of Fish Preservation

- ☆ Drying
- ☆ Salting
- ☆ Smoking
- ☆ Marinating
- ☆ Fermentation
- ☆ Combination of the above methods
- ☆ Though the methods are traditional, these are widely adopted in developing and under developed countries.

Advantages of Traditional Methods of Preservation

- ☆ Cheapest methods of preservation
- ☆ No expensive technology
- ☆ Can be employed in both small scale and large scale
- ☆ Used for the preservation of low value fishes
- ☆ Make the product available in the remote area as well as throughout the year
- ☆ Operational skill is not required
- ☆ No need of technically qualified people

2.1. Drying and Dehydration

Drying is one of the least expensive and oldest known methods of fish preservation. Drying under controlled condition is called dehydration. Though the technology of food preservation and processing has undergone revolutionary changes over the years and several new products processed employing diverse techniques have made their firm presence in the market, drying still continues to be the most widely used method for preservation of several foods including fish.

2.1.1. Drying vs Dehydration

2.1.1.1. Drying

Drying is under the sun by utilizing the atmospheric conditions like temperature, humidity and air flow. This is also called as sun drying. This involves removal of water from a body, in our context fish. Traditionally, fish used to be dried under sun and the terms 'drying' has come to imply drying under the sun. Sun drying is carried out in the open air using the solar energy to evaporate the water in the food. The evaporated water is carried away by the natural air currents. The efficiency of the process and the quality of the product remain at the mercy of the elements of nature and, therefore, the product suffers from serious disadvantages, some of which are:

- dependence of weather; the operations can be carried out only when bright sun light is available.
- long duration of drying; under unfavourable conditions of weather drying may take several days to complete.
- there is no control over the operating parameters
- possibility of contamination with dust, sand *etc.*
- possibility of infestation with insects, their eggs and larvae
- poor quality of the product
- short shelf life

However, it must be emphasised that the process of drying employed might have met the needs of the people of the time for preserved foods to sustain them during periods of derivations.

2.1.1.2. Dehydration

One of the aims of preservation and processing a perishable food is to avoid the wastage at the time of plenty and to make it available at a later time when its availability is limited or it is totally non-available. However, while achieving preservation care has to be taken to ensure that the wholesomeness, taste, texture, physical appearance *etc.* are also maintained to the maximum possible extent. Over the years several improvements have been brought about in the process of drying keeping in view the above consumer requirements. An important achievement in drying in this direction is the development of artificial driers where the important operational parameters like temperature, relative humidity and velocity of air *etc.* can be controlled. This has led to 'dehydration' which refers to a process of drying under controlled operational parameters like temperature, air velocity and relative humidity. In order that these carried out in an enclosed atmosphere. Therefore, in 'dehydration' the process is carried out in an enclosure provided with facilities to control the operational parameters. Most of the disadvantages encountered in sun drying can, more or less, be overcome and a product of desired quality and reasonable shelf life can be obtained by dehydration. However, the terms 'drying' and 'dehydration' are being used now a day without much specificity.

2.1.1.3. Advantages of Dried Foods

- ☆ Dried and dehydrated foods are highly concentrated foods compared to any other preserved form of foods.
- ☆ Drying reduces the microbial activity and thus reduces the spoilage of foods due to microbial activity
- ☆ With reduced water content enzymatic and many chemical processes are retarded
- ☆ Dried foods are less expensive to produce
- ☆ There is no involvement of complicated machinery and equipment for processing and packaging
- ☆ They are stable at most ambient temperatures
- ☆ Distribution costs are minimum

2.1.2. Principles of Drying

Water is essential for the activity of all living organisms including microbes. Reduction in the water content or its complete removal will proportionately retard or totally stop all microbial and autolytic activities, thus preventing spoilage and resulting in preservation.

2.1.2.1. Importance of Water Activity in Relation to Microbial Growth

Water content of fresh non-fatty fish is around 80 per cent; in fatty fish this may be less by a quantity equivalent to the percentage of fat. When the water content is reduced to a level below 25 per cent bacterial action stops. Further reduction to below 15 per cent can prevent mould growth. However, it is a common experience that dried fish containing relatively less water will spoil more quickly due to the action of bacteria than a similar sample containing more water when the latter has been salted before drying. Therefore, the expression 'water content' alone cannot satisfactorily explain the survival and growth of microorganisms in dried products. The relatively less growth of microorganisms in a substrate with higher water content depends upon its composition.

2.1.2.2. Water Content vs Water Activity (a_w)

The bonds between water and protein molecules in fish may be in several forms. Three layers of water, an adsorption layer, a diffusion layer and a free layer, surround the surface of a colloidal particle. Water at the adsorption layer that is tightly bound to the colloidal particle is called 'bound water'. The diffusion layer is less tightly bound to the adsorption layer. The farther it is from the adsorption layer, the more it behaves like free water. The third layer, the 'free water' has all the properties of ordinary water and can enter into all normal chemical reactions and support microbial spoilage. It is this 'free water', *i.e.* the water available to support microbial growth and react chemically, that is important in the drying process. The most effective physical method of assessing the state of water in foods is the measurement of partial vapour pressure of water in equilibrium with a given moisture content at constant temperature. This is called water activity, a_w. The a_w is a measure of water available to support microbial growth and chemical reactions. It is an important determinant of microbial growth and metabolic activity as well as of food deteriorating enzymes and chemical activity. Majority of food preservation procedures employs modification in the state of water in the foods. Water activity is expressed as the ratio between the water vapour pressure on the surface of the product and the water vapour pressure of distilled water, both at the same temperature.

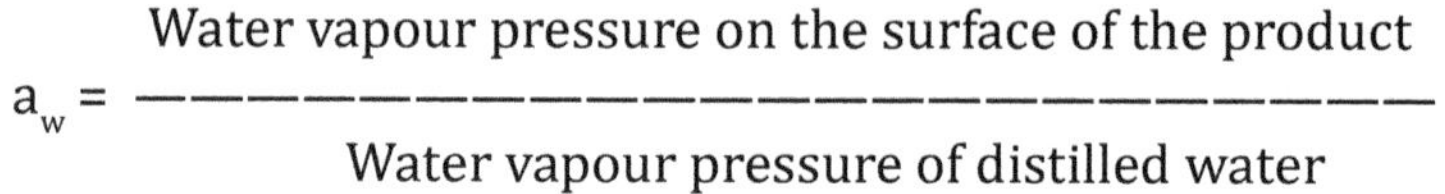

$$a_w = \frac{\text{Water vapour pressure on the surface of the product}}{\text{Water vapour pressure of distilled water}}$$

Water activity of pure or distilled water is assigned the value 1.0 and of an absolutely dry product 'zero'. Water activity in a food is expressed as a fraction relative to pure water. The a_w of salted fish will be low because of the presence of dissolved salt and this will explain why salted fish with relatively high moisture content will be more stable whereas unsalted fish with lower moisture content will spoil faster.

2.1.2.3. Water Activity and Related Microbial Growth

Fresh fish has a water activity above 0.95 and as it is lowered, microbial activity becomes gradually inhibited.

2.1.2.4. Measurement of a_w

One of the simplest methods of measurement of a_w of a food is in terms of its vapour pressure manifested as per cent RH generated in equilibrium with the food in a closed system at constant temperature. a_w is equal to the equilibrium RH divided by 100. A sample of food of known moisture content is allowed to equilibrate with a small headspace in a tight enclosure and then the equilibrium relative humidity (erh) is measured. The a_w of food = erh (air) ÷100.

Table 2.1: Water Activity (a_w) and Growth of Microorganisms

a_w	*Microorganisms Inhibited*
1	None
0.95	Gram-negative rods
0.91	Most spoilage bacteria – cocci, lactobacilli *etc.*
0.88	Most yeasts
0.80	Most moulds, Staphylococcus aureus
0.75	Most halophilic bacteria
0.65	Xerophilic moulds
0.60	Osmophilic yeasts

2.1.3. Drying Process

Drying process involves essentially two steps: migration of water from the wet parts of the materials to a vapourisation zone where the water vapurises and its removal from there to the surrounding atmosphere by the circulating air. These processes are called diffusion and evaporation respectively. The efficiency of a dehydration system will depend on how these factors are controlled.

2.1.3.1. Factors Influencing Dehydration

i. Equilibrium Moisture Content and the Influence of Humidity

The pressure of water vapour at the surface of the fish is different from that of water in the air. Because of this difference in partial pressures migration of water molecules takes place from one medium to the other. The partial pressure of water vapour at the surface of fresh fish, which is saturated with water, is greater than that of the water vapour in the air. Therefore, migration of water vapour from fish to the surrounding air, *i.e.* dehydration, takes place. The converse will be true if partial pressure of water vapour at the fish surface happens to be less. The processes of sorption (uptake of water) or desorption (release of water resulting in dehydration) will continue until the pressure of water vapour at the fish surface and in the air is equalised. The moisture content in the fish is then in equilibrium with the air humidity. The moisture content, which is in equilibrium with the atmospheric humidity, is called the equilibrium moisture content.

Equilibrium moisture content is a very significant factor in a drying operation, particularly in a controlled operation. Equilibrium moisture content of fish under drying will vary depending on the variations in the vapour content in the air. Any method of drying by exposure to air can remove only that moisture in the fish, which is over and above the equilibrium moisture content possible at the operating air temperature and humidity. Under a given set of conditions the drying rate can be increased only by affecting a shift in the equilibrium moisture content by dehumidifying the air in the drying atmosphere.

ii. Heat Transfer

Drying of fish involves the removal of water from the solid fish body with the help of an input of thermal energy. The outward flow of water vapour is linked with an inward flow of heat. Thus drying is linked with the rate of supply of heat to the point from where evaporation occurs and its conduction within the material. Transfer of heat by conduction takes place by the exchange of momentum of random thermal motion of the molecules to the neighbouring molecules, *i.e.* passed along hand to hand. However, the heat transfer should also consider provisions of latent heat of evaporation.

iii. Mass Transfer

As has been mentioned earlier drying is a phenomenon involving migration of water within the body to the surface and conveyance of vapourised water away from the surface. The factors controlling the rate of transfer of water from characteristics of the body's environment, especially the temperature, pressure, humidity and air velocity. The factors, which determine the rate of movement of water within the body, are mostly independent of the external conditions.

Some of the physical mechanisms which are considered as involved in mass transfer are:

- liquid movement under capillary force
- diffusion of water caused by a difference in concentration
- surface diffusion in liquid layers adsorbed at solid interfaces
- water vapour diffusion in air filled pores caused by a difference in partial pressure
- water vapour flow under differences in total pressure as, for example, in the vacuum drying under radiation
- water flow caused by shrinkage and pressure gradients
- flow caused by gravity
- flow caused by vaporisation-condensation sequences.

2.1.4. Factors Influencing Drying Rate of Fish

Rate of drying is influenced by several factors-the intrinsic nature of the fish being dried and several external factors employed in the drying atmosphere.

2.1.4.1. Nature of the Material being Dried

The individual nature of the fish, its chemical composition and physical structure and important factors influencing the rate of drying. Under constant external conditions higher rate of drying will be exhibited by the fish.

- Containing more water

☆ Having larger surface area in relation to weight. Splitting the fish will result in increase of surface area relative to weight and the rate of drying will become faster.

However, the presence of high mounts of fat will retard the rate of escape of water and hence, the rate of drying.

2.1.4.2. Relative Humidity (RH) of the Air

Absolute humidity (H) is the mass of water contained by a unit mass of air.

$$H = \frac{\text{Water Vapour (kg)}}{\text{Dry air (kg)}}$$

Relative humidity (RH) is the absolute humidity divided by the humidity of saturated air under identical conditions of temperature and pressure assuming unit mass of dry air in both cases. In other words, relative humidity refers to the degree of saturation of vapour space or body of moist air and is expressed in per cent saturation of air with moisture at a given temperature. The most important single factor correlated with drying rate is the relative humidity of the drying air passing over the fish under drying. RH is measured in terms of wet bulb depression. Wet bulb depression is the difference between the temperature readings of two thermometers kept in an air current, one with its bulb kept damp (wet bulb) using a wetted muslin sleeve and the other uncovered and dry (dry bulb). In the initial stages of drying this corresponds to the difference between the temperature of wet fish surface (Wet bulb) and that of the drying ait. The difference between dry bulb and wet bulb temperatures is related to the RH. When the difference becomes zero it implies that the air is saturated with water vapour (RH 100 per cent) and it cannot carry any more water and hence the fish will not dry any further. As drying progresses in a closed system eventually such a situation arises. A change in air temperature is usually accompanied by a change in RH. By increasing the air temperature its moisture carrying capacity can be increased; however, this may adversely affect the quality of the product.

During the later stages of drying, the rate of internal diffusion of moisture becomes less than the rate at which it is evaporated from the surface. At this stage it is desirable to maintain the RH of the air at a relatively higher level in order to avoid over-drying of the fish surface and thus to avoid or minimise the occurrence of case hardening.

2.1.4.3. Air Temperature

In the beginning of the drying cycle evaporation of water produces a cooling effect on fish and air around and hence the temperature of the fish tends to fall below the ambient. Difference in temperature does not show any appreciable influence on the drying rate at this stage. Neither the wet bulb depression not air velocity significantly affects the drying rate. To maintain constant temperature of drying the heat used for evaporation has to be fed back to the system. For every kg of water to be evaporated 550 kcal (2600 kJ) of heat must be supplied to the system.

The influence of temperature on drying rate becomes more pronounced in the later stages of drying when the moisture content falls. In the low moisture range the drying rate is so slow that the cooling effect of evaporation is not appreciable. The fish being dried attains very nearly the temperature of the drying air. The internal redistribution of water is the true factor which determines the drying rate in this phase. This can be accelerated by a rise in the materials temperature.

The recommended temperatures for drying fish of temperate waters is 25 to 35°C whereas tropical fish is generally dried at 40-50°C. Higher temperature may cause cooking of fish flesh which will make the resultant dried product brittle. Therefore it is important that the temperature employed at any stage of drying is not high enough to result in cooking of fish flesh.

2.1.4.4. Air Velocity

The rate of drying will be faster with higher velocity of air passing over the fish. In a drying system three layers of air can be considered to be prevailing over the fish – a stationary layer close to it, a slowly moving layer outside this and a third outer turbulent layer of air. The stationary layer next to the fish becomes saturated with moisture, which passes into the slowly moving layer. When the air velocity of the outer layer is high, the slow moving air will become thinner allowing more rapid movement of water away from the fish surface.

The air velocity should be maintained such that the processes of diffusion and evaporation proceed smoothly. Air velocity above a certain range does not have any beneficial effect on the rate of drying. In the later stages of drying when rate of diffusion of moisture becomes less than the rate at which it evaporates from the surface, a lower air velocity at a level which permits adequate circulation and heat transfer will be desirable.

At an air velocity of 70 m/min the drying rate is twice as rapid as in still ait. It is three times as rapid as in still air when the air velocity employed is 140 m/min. It is considered best to employ an air velocity between 75 and 130 m/min for fish drying.

2.1.4.5. Constant and Falling Rate Drying

Non-hygroscopic solids, in general, exhibit two distinct phases of drying; the constant rate drying and the falling rate drying.

i. Constant Rate Drying Period

In the beginning of drying the fish surface is saturated with water and evaporation takes place as from a free water surface. The rate of escape of moisture per unit weight per unit time will remain a constant aslong as the surface remains wet. This is governed by conditions of the surrounding air like temperature, RH and velocity which will influence how rapidly the air can supply heat to the water in the fish and remove the vapour produced. During the pendency of this period heat absorbed for evaporation of water at the moist surface is balanced by heat flow from the warm air stream into the cool wet body and both are in equilibrium. In physical terms water is diffusing to the surface of the fish at a rate equal to its removal from the surface. The temperature of the fish being dried, generally, is the wet bulb temperature of the air in contact with it.

ii. Falling Rate Drying Period

At a stage in the drying process, water can no longer diffuse to the surface, as rapidly as it is evaporated and, therefore, the surface no longer remains wet. The moisture content at this stage called the "Critical moisture content". The rate of drying then will depend on the rate at which water in the fish migrates to its surface. The zone of evaporation recedes deeper and the rate of movement of moisture continues to decrease; the drying rate falls progressively. The rate of drying is controlled by the rate of diffusion of moisture which is influenced by several factors.

- ☆ Composition of fish – high fat content retards the rate of drying.
- ☆ Thickness of fish – with thicker fish water in the middle layers has to travel long to reach the surface.
- ☆ Temperature of the fish –diffusion of moisture from the deeper layers to the surface will be greater at higher temperatures. This is the only aspect which can be controlled to derive beneficial effects during the falling rate period.
- ☆ Moisture content – when the moisture content becomes low the rate of movement to the surface is reduced.
- ☆ Amount of added salt – the higher the amount of added salt, the more slowly the moisture diffuses to the surface.

However, the drying rate at this phase does not depend on the humidity of the air, provided the air is not saturated. The drying rate also does not depend on the speed of the air passing over the fish. This phase of drying is called the falling rate period. The solid materials in the fish begin to absorb heat from the air and the temperature of the fish begins to approach the dry bulb temperature of the air.

In many cases the fish is dried after salting when a substantial quantity of body water is lost through osmosis. Salted fish, therefore, does not exhibit a constant rate drying period. A continuously failing drying rate is exhibited by such fish during dehydration.

2.1.4.6. Effect of RH of Air of the Final Water Content of Fish

During drying under a given set of operational conditions the loss of moisture stops at some point and it becomes impossible to remove any further water from fish. Relative huminity of the air in the drying atmosphere is the factor controlling this phenomenon.

Table 2.2: Relation between RH and Minimum Moisture Content Obtainable in Lean Fish

RH of the Air (Per cent)	*Minimum Moisture Content Obtainable in Fish (Per cent)*
20	7
30	8
40	10
50	12
60	15
70	18
80	24

2.1.5. Fish Drying Process

2.1.5.1. Pre-process Operations

Before drying, the fish is often subjected to different pre-process operations. These are mainly dependent on the size and nature of the fish, as also the end product desired.

i. Cutting/Splitting

Conventionally, very small and thin fishes like anchoviella are dried whole without salting. Small fishes like sole, small croakers, anchoviella *etc.* are salted whole and dried. However, gutting and cleaning before salting will reduce spoilage and improve the quality. Splitting open or cutting into pieces before salting is necessary in the case of bigger fishes. In the case of very big cat fish, shark, rays *etc.* they are split open and deep scores are made in the exposed flesh before salting. This will increase the area of exposed flesh for greater contact with salt, permit easy penetration of salt and increase the surface area for evaporation of moisture, thus bringing about a reduction in the time required for drying.

ii. Salting

Fish may be dried without salting or after salting; it is better that the fish is immersed in brine rather than in dry salt so that the required concentration of salt in fish flesh is achieved in a shorter time. When fatty fish is salted, air should be excluded during brining to reduce the rancidity. It is also desirable to employ saturated brine to enable easier uptake of salt by the fish flesh.

2.1.5.2. Natural Drying

Solar and wind energies are made use of in natural drying process. Natural drying has been known to yield very stable dry fish and several such products are known to have remained safe for years together. Some of the essential requirements for production of high quality dried fish by natural drying in the tropics are:

- ☆ sufficiently high air temperature. Temperature in the range 35-40°C will be ideal. In many tropical countries temperature often becomes higher.
- ☆ sufficiently low RH to permit drying fish to an a_w level not conducive for bacterial spoilage. RH above 70-75 per cent will not help dry the fish to the desired level. Salted fish will tend to absorb moisture from the surrounding air if the RH rises above 75 per cent. In coastal regions the humidity will be often very high limiting the speed of drying.
- ☆ use of raised platforms. Air movement at ground levels is comparatively slow. Better air movement can be ensured if the fish is kept raised by about one metre above the ground.
- ☆ use of drying racks. Keeping the fish on racks kept above ground level will facilitate movement of air both under and over the fish, thus allowing drying of fish by dust or sand also will be minimised. Racks with sloping tops will allow for easy draining of any surplus water on fish surfaces in the beginning of drying.

2.1.5.3. Artificial/Mechanical Drying

Mechanical driers can be broadly classified into two types. In one type, the heat is transferred into the product through a hot gas, usually air. A heat/water vapour exchange takes place at the contact point of hot air with the product and the outgoing air will carry the water vapour away. Kiln drier, tunnel drier *etc.* are examples of this class of driers.

In the second type, heat is transferred to the product through a solid surface which may also be used as the carrier for the product to be dried. A typical case is a vacuum shelf drier where the hollow shelves serve both as heat transfer medium and also hole the material to be dried. The whole drying cabinet is evacuated and the water vapour produced is removed using a vacuum pump. However, in all such types of drying the product need not necessary be held under vacuum, it can be exposed to air as well.

Advantages

1. It is the cheapest fish preservation method among all other methods.
2. It is a very simple process, can be adopted by anybody even an illiterate person.
3. The process involves simple implements, little/no machineries, very little investment, being labour intensive.
4. This is the method, in which most of the body materials is retained, from nutritional point of view, sun dried fish is superior without loss of an y nutrient.

Disadvantages

1. Excess drying results in charring which reduces the flavour and nutrient.
2. This method of preservation results in great reduction in weight of the fish, the yield by drying is on an average 25 per cent as such cost of finished product *i.e.* dry fish, is always higher than the raw materials.
3. Self life of dry fish (with out salting) ranges from only 2-6 months in prime condition.

Table 2.3: Improvement in Fish Dry Methods

Sl.No.	*Drying over Ground*	*Improved Drying Method over Racks*
1.	Wind is very weak at ground level	Wind is stronger above the ground level
2.	Air passes only on one side of the fish other side being in contact of ground	Air passes over all sides of the fish for efficient drying through dehydration
3.	Fish dried on ground is prone to infection by sand,dirt,insects and pests	Fish dried above the racks above the ground level is hygienic and attractive
4.	Hot sand can cook and burn the fish making it fragile	Fish dried over racks will not cook or burn, all sides being well ventillated
5.	During rain fish on the ground even if covered, gets wet through the moisture from the soil.	During rain fish on rack can be covered and protected from hydration through the rain water.

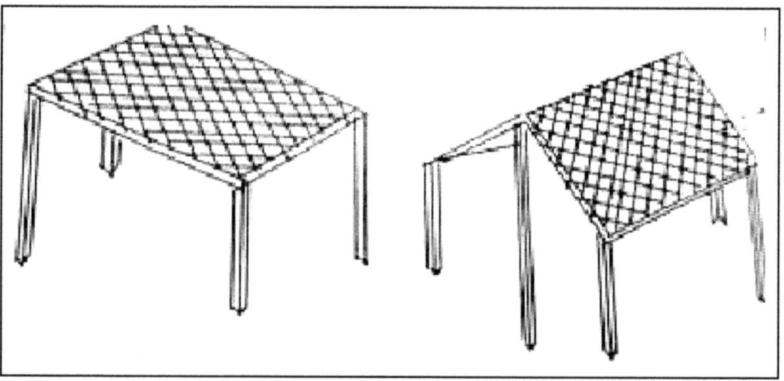

Figure 2.1. Fish Drying in Raised Platforms.

2.1.5.4. Cabinet Drier

This is a simple batch operation model drier used for relatively small scale operations. A typical cabinet drier may consist of an insulated or non-insulated framed structure. Materials to be dried, uniformly spread in trays may be placed on permanent supports provided in the drier. A fan located inside the drier will blow air past a heat source which pass across or through the materials loaded in tray.

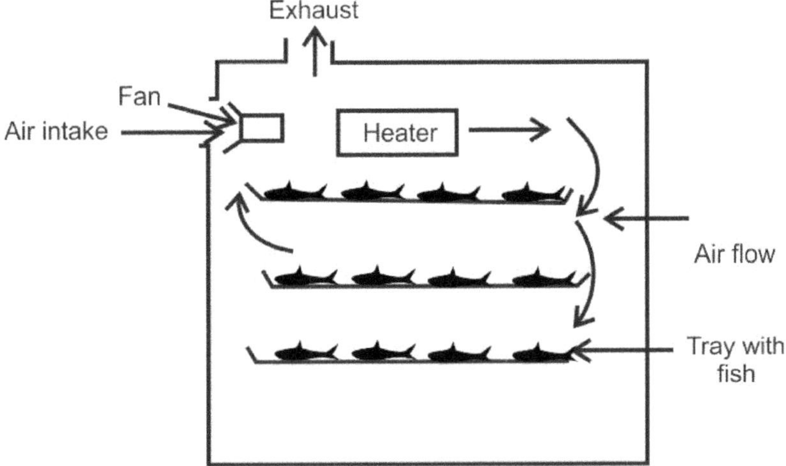

Figure 2.2. Cabinet Drier

2.1.5.5. Kiln Drier

Kiln drier is a batch drier. A typical drying kiln will consist of a two-story building. The floor of the upper story is slotted or may be composed of narrow slats on which the materials can be spread. This story serves as drying room. The burners or furnace producing hot gas is located in the lower floor. The hot gas passes through the product by natural conduction; often forced circulation with the help of a fan also may become necessary. The material being dried has to be turned and stirred frequently to ensure uniform drying.

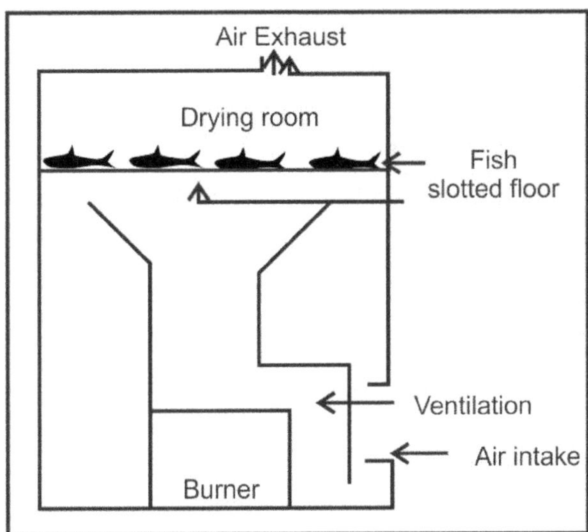

Figure 2.3. Kiln Drier.

2.1.5.6. Tunnel Drier

Tunnel driers are most commonly used for drying fish. These are made in the form of long tunnels, 10-15 m long. Trolleys loaded with trays containing fish are moved at a predetermined schedule through the tunnel. Hot air is blown over the material across the trays. For drying fish using the tunnel driers the production cycles are so planned that when a trolley of fresh fish leaves the drier at one end, a fresh trolley of fish to be dried is introduced at the other end.

i. Parallel Flow Drying Tunnel

In tunnel drying the movement of air can be maintained in either direction in relation to the flow of the material. In some systems the air movement is in the same direction as that of the product movement. This type of drier is called a parallel flow drying tunnel. In such driers the hottest air comes in contact with the wettest fish. Therefore relatively higher temperature can be employed for drying. One serious disadvantage of the system is that the air towards the outlet side will become cool and highly humid; therefore the finished product may not be sufficiently dried.

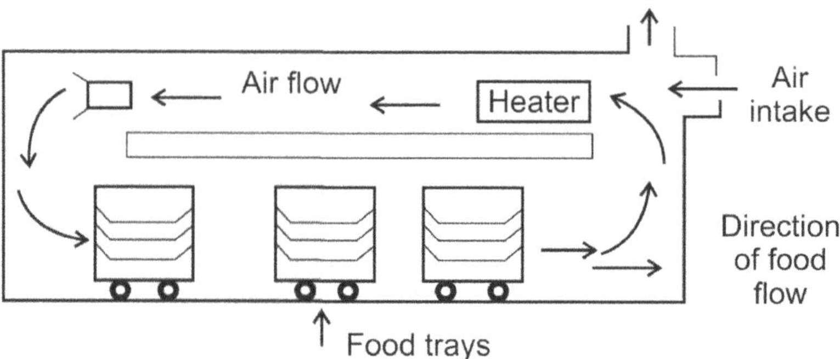

Figure 2.4. Parallel Flow Tunnel Drier.

ii. Counter Flow Drying Tunnel

The movement of air in this type drier is in the direction opposite to the movement of the product. The hot dry air comes first in contact with the driest materials so that the finished product obtainable will be very dry. However, in prolonged drying schedule the fish from the other end of the tunnel may remain in humid warm air for long periods without getting sufficiently dried. Bacterial and other types of spoilage are likely to take place in the fish during this period. Dried product also should not be left to remain in the drier for longer periods than necessary. Otherwise, the product may become too dry and will affect the organoleptic characteristics.

Tunnel dries may be designed based on the concept of hot air recirculation or allowing the hot air passing over the fish to escape to the atmosphere. The recirculating air will become highly humid to the extent of slowing down the drying

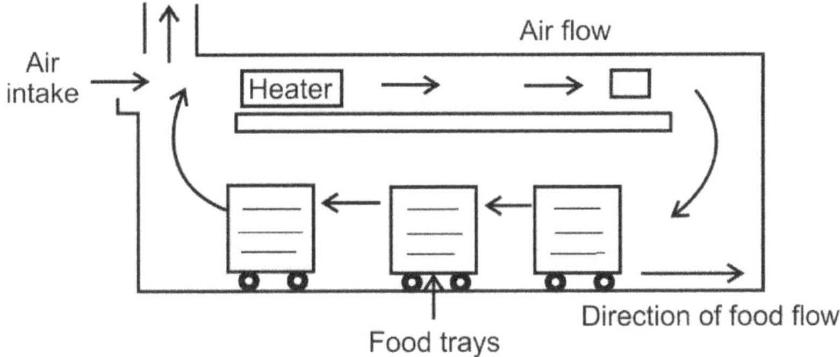

Figure 2.5. Counter Flow Tunnel Drier.

significantly. Therefore, it is very essential that tunnel driers with provisions for recirculating hot air should have adequate facilities for dehumidification and control of humidity in addition to control of air temperature.

2.1.5.7. Spray Drier

Spray driers are generally used for drying foods which are in the form of liquids or suspensions. In principle a food in a liquids or paste form is atomised and dispersed as minute droplets which are suspended in a stream of hot air in a chamber where it gets rapidly dried. The dry particles suspended in the air stream flow in to separation equipment where they are separated from the air, collected and packaged. In the application of spray drying to fish products it is limited to products like fish protein hydrolysates and fish powders.

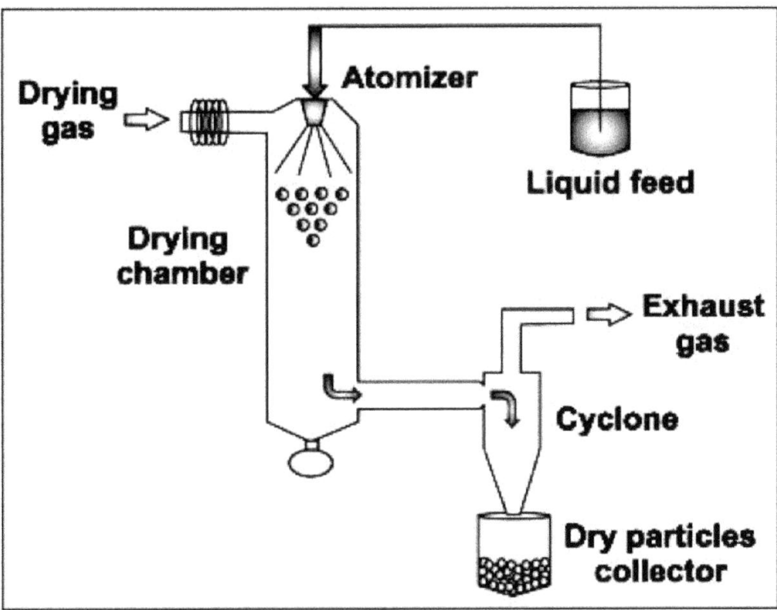

Figure 2.6. Spray Drier.

Better retention of colour, flavour and nutritive value are the advantages associated with spray dried foods.

2.1.5.8. Heat Transfer Through Solid Surface

i. Drum Drier

Drum drier is an example where heat transfer to the product takes place though a solid surface. Drum drier is used for drying fluid materials. The food product in the form of slurry is deposited as a thin film on the drum. The drum is heated, generally by steam; while it is being rotated. Drying can be done keeping the drum open to the atmosphere. If the material is desired to be dried under vacuum, the drier can be enclosed in a vacuumed chamber. The product when dry is removed from the surface of the drum using a scraper blade.

Drum driers are classified as single drum, double drum and twin drum types. Single drum drier comprises only one roll. Double drum comprises two drums rotating towards each other. Twin drum is similar to double drum, but rotate away from each other.

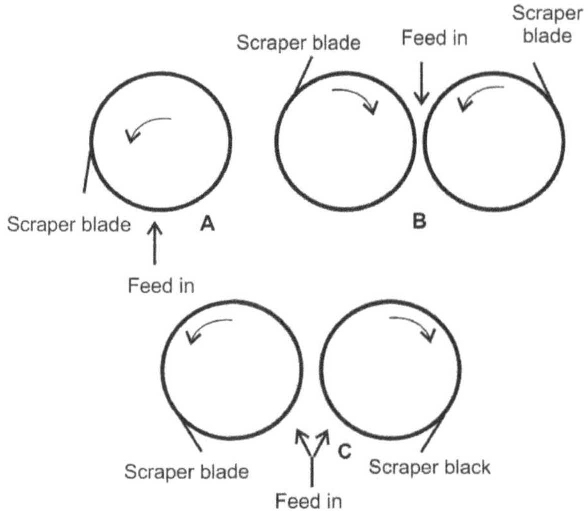

Figure 2.7. Types of Drum Drier: (A) Single Drum; (B) Double Drum; (C) Twin Drum.

ii. Vacuum Shelf Drier

Vacuum shelf drier consists of a vacuum tight chamber of heavy construction with access door and outlet for gases and vapours. Hollow shelves, through which the heating medium is circulated, are fitted inside the chamber. The materials to be dried is spread in fairly thin layers in metal trays which rest on these shelves. Alternately the materials can be spread directly on the shelves. Heating is done by circulating hot oil, steam or any other suitable heating medium. Vacuum will be drawn in the chamber through the vapour outlet, and drying will proceed under vacuum. The initial drying rate will be high; however, as the drying proceeds the material with shrink and tend to curve away from the trays. This will reduce the effective area of contact of the material with the heating surface which will cause a decline of heat transfer to the material thus slowing down the drying rate.

Vacuum drying is considered an expensive process. However, it is quite suitable for drying fatty fishes where the probability of fat oxidation and rancidity in the product can be minimised.

2.1.6. Solar Drying

Harnessing the solar radiation for solar powered driers has, of late, attracted considerable interest because of the absence of any energy cost and possibility of producing a dry fish in god hygienic condition even when the RH is high. Energy of the sun is collected and concentrated to produce elevated temperatures suitable for drying several commodities including fish. When the temperature of the air is raised, its RH will be reduced, in other terms its capacity to hold water will increase and hence, can absorb additional quantities of vapour.

2.1.6.1. Solar Tent Drier

One of the simplest forms of driers to use solar energy for drying fish is the solar tent drier. This is working on the principle that a black surface absorbs sun's energy far more effectively than any light coloured surface. The air thus heated is allowed to pass through the fish and escape out through a vent in the top simultaneously admitting fresh air inside through a vent provided at the bottom of the tent.

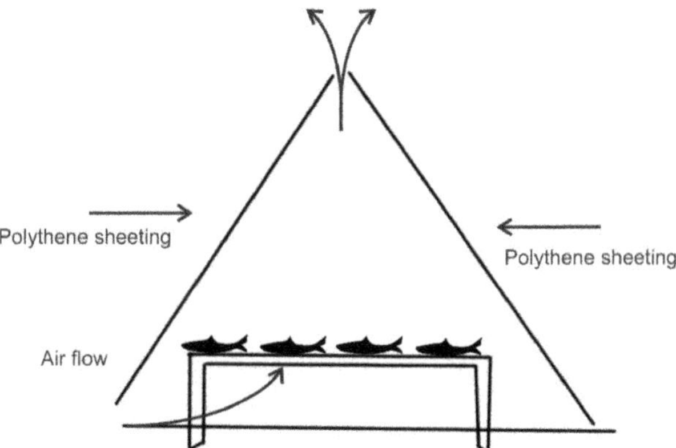

Figure 2.8. Polythene Tent Drier.

In a solar tent drier, the air temperature is known to rise to the levels of 60°C or more in tropical climate. This will adversely affect the nutritional as wwell as physical properties of fish and is considered a disadvantage. However, compared to normal sun drying solar drying has the following advantages:

- No energy cost
- Very low equipment cost
- Shorter drying periods
- No contamination from dust, insects
- Produces hygienic product with low moisture content

i. Other Types of Solar Driers

The tent drier described in section 2.1.6.1 above is perhaps the simplest in design and operation. Several attempts have been made to improve the efficiency of the drying operation, and accordingly, several designs have come up. Many of these have not become popular due to some reason or other. However, solar drying for some other commodities like food grains have become a commercial success.

2.1.7. Effect of Drying on the Quality of Fish

2.1.7.1. Shrinkage

The major apparent physical change which can be observed in dried fish is the shrinkage in its volume. Shrinkage takes place as a result of the structural changes taking place in the fish muscle during drying. Ordinary drying is liquid phase drying and when water leaves its place, proportionate shrinkage in volume of fish also takes place.

2.1.7.2. Case Hardening

Water in the fish contains dissolved salts, proteins and other organic matters. Water moving to the surface of the fish carries all these, while that leaving the fish surface is only pure water depositing the dissolved substances on the surface. If the temperature of drying air is high and its RH is low this will form a dry impervious layer on the surface preventing further diffusion of water. This condition is referred to as 'case hardening'. When case hardening takes place, the temperature of the fish muscle will increase resulting in its cooking. This will result in the final product becoming brittle. Case hardening greatly affects the rehydration property of dried fish very adversely.

Case hardening can be controlled by maintaining sufficiently high RH in the drying atmosphere as also controlling temperature of drying.

2.1.7.3. Denaturation of Protein and Toughening of Texture

As the drying progresses, concentration of the dissolved material in the body water increases. Reduced evaporation due to case hardening will result in increase of temperature of the fish muscle. These bring about denaturation of protein and the texture becomes tough besides loss of juiciness of the meat.

2.1.7.4. Rehydration

Air-dried fish is hard in texture. The proteins are denatured and the water-holding capacity of the proteins become irreversibly lost. Therefore, penetration of water will be greatly retarded resulting in poor rehydration.

2.1.7.5. Effect on Colour, Flavour

Pigments, fat *etc.* in fish are susceptible to oxidation. Prolonged drying of fish exposed to hot air can accelerate this process. Therefore air-dried fish often suffers from discoloration and rancid odours and flavours. Another type of discolouration is the on-enzymatic browning developed through Maillard reaction. Free sugars react with free amino groups and produce brown coloured products in fish.

2.1.8. Spoilage of Fish during Drying and Storage

Dried and drying fish are susceptible to many types of spoilage which can affect the quality and shelf life. Damages occurring due to flies and insects are of great significance in open drying under the sun. These are, to a great extent, controlled in mechanical drying, but the post-process infection as well as damages taking place during storage are of equal importance in either type of drying. Most of these can be controlled by adopting hygienic practices in the production side as also maintaining appropriate storage atmosphere.

2.1.8.1. Moulds

Moulds can grow on the salted or unsalted dried fish if the moisture content is high and high RH above 75 per cent prevails in the storage. Moulds are likely to grow if the temperature of storage is maintained in the range 30-35°C. with the onset of moulds the surface moisture may increase and the fish may become susceptible to other types of spoilage.

2.1.8.2. Insect Infestation

Unsalted dried fish are often infested with blowflies, *Chrysomya* spp., *Lucilia* spp., *Sarcophaga* spp, *etc.* Adult flies are attracted to the wet fish and their larvae feed on it.

Small fish may dry quickly and hence the larvae may perish before highly damaging the fish. However, larger fish may remain wet for longer periods and, therefore, the incidence of damage due to flies will be more in large fish.

Infestation with flies can be largely reduced by maintaining hygienic conditions in fish handling and processing premises, eliminating areas where larvae can pupate and adult flies lay eggs, and using salt in the drying process.

Insect attack takes place in dried fish during storage as well. Beetle infestation is a serious problem in dry fish during storage. *Dermestes* spp. are the predominant blow flies associated with this. They can grow at moisture levels around 15 per cent. Their larvae feed on the fish, often leaving only bones.

Although salting does not result in killing the larvae or adult flies, presence of salt can reduce their activity.

2.1.8.3. Rancidity

Fatty fish, particularly, are prone to oxidation and development of consequent rancid flavour. Though some degree of rancid flavour may be acceptable in dried fish, excessive rancidity will be objectionable. Randicity can be controlled to some extent with airtight packaging of dried fish.

2.1.9. Packaging

Dry fish poses some problems with respect to their packaging needs.

- ✰ The irregular shape, whether dried whole or as pieces;
- ✰ Sharp protrusions which may puncture the packaging materials and occupy space;
- ✰ Dried fish often may be brittle and will break up when packed in a regular shaped package;
- ✰ Individual packaging for fish will shoot up the cost and
- ✰ Dried fish, if not appropriately protected by package will absorb moisture and spoil quickly, necessitating water-proof/vapour-proof packaging.
- ✰ An ideal package for dry fish should have the following desirable properties:
- ✰ Be resistant to mechanical abrasion and puncture. To reduce the breakage during handling and distribution, the package should be sufficiently robust to retain its integrity. Containers of corrugated fibreboard will be good as they can provide adequate mechanical protection and
- ✰ Be impermeable to moisture, oxygen, and insects. Boxes made of corrugated fibreboard or polypropylene can meet these requirements if they are also sealed throughout.

However, it remains a fact that practically there is no proven solution to all the technical problems in packaging of dried fish. Any improvement in the existing system should be appropriate to the system, particularly in economic terms.

2.2. Salting/Salt Curing

Salting is a traditional method of preservation of fish, practised as such or in combination with drying or smoking, and is considered a practice as old as drying. Salt curing, though is an important method of preservation by itself, with or without subsequent drying is one of the most widely practised methods of fish preservation throughout the world. When introduced in sufficient quantities in the fish flesh, salt can delay the activity of bacteria or even inactivate them by reducing the water activity. This forms the basis of preservation by salting.

The advantage of salt curing over many other methods of processing are:

- ✰ Does not require elaborate equipment;
- ✰ Capital outlay is small;

- ☆ Methods are simple and the processes are comparatively inexpensive;
- ☆ Unlike other sophisticated methods of processing, curing can be applied for preservation of any type of fish;
- ☆ The finished products do not require any special storage facility;
- ☆ Nutritionally the products are comparable to fish processed employing most other methods.

2.2.1. Salting and Related Water Activity (a_w) of the Fish Flesh

When fish is mixed with salt or kept in salt solution, some water is removed from the flesh depending on the amount of salt used, dry or in solution. Loss of water from fish flesh reduces its a_w. During salting process salt enters the fish flesh and water moves out due to osmosis. The method of salting is so selected that the salt penetration is rapid enough to similarly lower the water activity in the deepest part of the fish. This process will continue until an equilibrium is attained between the concentration of salt in the fish muscle and the surrounding brine. Water activity of saturated brine is 0.75. Fish flesh in equilibrium with saturated brine also will have a_w similar to this, say 0.75.

2.2.1.1. Salt Concentration and the Bacterial Growth

Presence of salt at 4-10 per cent level in the fish flesh is known to prevent the action of most spoilage bacteria as well as autolytic decomposition. In general, growth of many putrefactive bacteria is considered controlled with salt concentration over five per cent. However, there are several other bacterial groups which can tolerate higher concentrations of salt. Therefore, for fish curing higher concentrations of salt are always used. When the concentration of salt is 20 per cent or more in the flesh the decomposition process in the fish proceeds only very slowly.

i. Salt Tolerance and Bacterial Groups

With reference to the sensitivity towards salt, bacteria can be divided in to three groups:

- ☆ **Halophobic or Salt Sensitive:** These include most of the pathogens and putrefactive types; *e.g. Pseudomonsas* and *Achromobacter* spp. These organisms cannot grow in a medium where the salt concentration is higher than six per cent.
- ☆ **Halotolerant:** These consist of spore-formers, micrococci and some anaerobes, particularly Clostridium butulinum. They can grow in concentrations higher than six per cent and even up to saturation (beyond 15 per cent), although the rate of growth decreases with increasing concentration salt.
- ☆ **Halophilic:** These include all salt loving organisms which grow best in the presence of salt. Practically, the halophiles fail to grow in the absence of salt and the optimum growth occurs in salt concentrations higher than

20 per cent. Bacteria causing 'red' and 'pink' in salted fish are the most important in this group. These are aerobes with optimum temperature of growth in the range 15-50°C. Most of them fail to grow below 10°C. They are not pathogenic.

Defects of Traditional Salt Curing

The main defects of the traditional method of salt curing are:

1. the fish are tainted by the time they ring is each the curing yard, nor are the fish gutted and cleaned after being caught.
2. drying is done on ground or over very thin mats as such lot of fine sand and clay gets into the flesh, due to the wind flow.
3. insufficient salt is used and that to for a very short time, as such drying is also incomplete.

2.2.1.2. Action of Salt

During salting, sodium chloride penetrates the fish tissues. While doing so, salt alters the colloidal properties of protein and changes the nature or hydration. Sodium chloride being strong electrolyte releases some of the bound water of the protein and this affects its state of existence. In low concentrations, salting actually results in swelling of the flesh whereas at high concentrations salting out takes place whereby the fish flesh in depleted of considerable amount or water. This will partially account for the hard texture of salt cured fish.

Sodium chloride exerts a high osmotic pressure giving rise to reduction in moisture content of the fish and plasmolysis in the bacterial cells as a consequence. It also blocks the protein unclei, which are affected by enzymes. It alters the state of proteins and enzymes in such a way that the protein becomes impervious to the actions of the enzymes. Enzymes are also destroyed or made inactive by the salt, thereby preventing autolytic degradation.

Sodium chloride forms brine using the water on the surface of the fish and penetrates the fish flesh by a dialysis mechanism. Water diffuses from the fish flesh tissues to outside due to osmotic pressure between brine and the fish muscle solutions. The process will continue until equilibrium attains between the salt concentrations in the fish flesh and the surrounding brine.

2.2.2. Salting Process

Salting, though a method of preservation by itself, is also resorted to as a preliminary to other methods of processing like smoking, canning *etc.* The salted fish may be expected to remain stable for a long time, though often it is intended only as a short-term preservation method. Therefore, the selection of method of salting depends primarily on the end product.

2.2.2.1. Preparation of Fish for Salting

Whatever be the ultimate product intended, for successful salting it is essential to ensure that the whole surface of the fish is in contact with salt. To achieve this the fish is dressed. Usually, the fills and entrails are removed and the fish is split open along the vertebral column. In the case of big fish, or they are filleted. However, when very small fish are salted, they are salted whole. Principally, it is the size of the fish that decides whether the fish should be salted whole, uneviscerated, eviscerated, split open or made into fillets.

2.2.2.2. Salting Methods

Dry salting, kench salting, brine salting and, mixed salting are the common methods employed for salting fish.

i. Dry Salting

Dry salting is the simplest method of curing fish. Dressed fish are kept intimately mixed with crystalline salt in containers. Salt may even be rubbed into the gill cavities and the scores made in the fish flesh. A layer of salt is spread on the bottom of the tub and a layer of fish is placed on it skin down. Salt and fish are spread in alternate layers, the proportion of salt increasing upwards. A solution of salt will be remain in this self-brine. The fish may float in the brine formed. In order to ensure proper salting of all the fish, weights are often used to keep them immersed in the brine. Fish will be allowed to remain in brine for two to three days after which it can be taken out and dried or otherwise disposed. Yield of the product by this method is about 35-40 per cent.The product has a shelf life of 6-10 weeks.

ii. Wet Salting

The initial stages of processing and salting are the same as dry salting. Once the fish is put into the tank it is allowed to remain in the self brine. The fish is not dried at all. The wet fish is then drained and packed in palmyrah leaf baskets or coconut leaf baskets and taken to the market. The fish is taken out only when there is demand. This method is particularly suitable for fatty fishes. This is mainly done for fishes like oil sardine, mackerel, ribbon fish *etc.* In such fishes the fat gets oxidized on exposure to air. These products have moisture content of 50-55 per cent and the salt content around 25 per cent. They are most susceptible to fungal attacks, bacterial degradation and general putrefaction. They have a very short shelf life.

iii. Kench Salting

Kench salting is essentially a method of dry salting except than the self-brine formed is allowed to drain off. Split fish, after rolling up in crystalline salt or rubbing salt onto the surface, is stacked in layers interspersed with a thin layer of salt between each layer to a height of 1 to 1.3 m. In the lower layers the fish is stacked flesh up and in the top layer it is kept skin up. The self-brine formed is allowed to drain away.

iv. Brine Salting

In this method of salting, fish is kept immersed in brine of desired concentration for the required time. This is usually done as a step preliminary to smoking and canning. A strong cure is not possible in brine salting because the brine will become progressively diluted with the water released from the fish. If a strong cure is needed the brine should be replaced with a strong one after the initial brine becomes diluted.

v. Mono Curing

Mono curing is mainly done on medium to small size fishes. The curing is done without splitting the fish. However, the intestine and entrails are removed by pulling out through the gill region. The fish is then salted and kept in tanks. The flesh is not exposed during salting thereby causing less contamination. The yield is about 70 per cent and product has shelf life of 50 days.

vi. Pit Curing

The fish is mixed with salt in the ratio 4:1 and put in pits dug on the beaches. The pits may be lines with palmyrah/coconut leaves. After 2-3 days the fish is taken out and packed in bamboo baskets and transported to markets without drying. The quality is poor and the fish is highly contaminated with sand and has shelf life of about 20 days. This is commonly called as "Kuzhi Karuvadu".

vii. Colombo Curing

Colombo curing is actually a pickling process for wet salt curing of pelagic fish. Colombo curing was practiced by the fishermen of South Canara and Malabar regions of the west coast of India. In the past, Colombo-cured mackerel and sardines were exported from India to Sri Lanka in large quantities. This method has now become obsolete after the advent of freezing and canning for fish processing and the widespread use of ice. Mackerel and sardines, available in huge quantities during the rainy season, were chiefly used for Colombo-cured fish is sour in taste and fibrous in texture. Several years ago, Sri Lanka stopped the import of Colombo-cured fish from India. All these factors have contributed to the disappearance of this once well known process. Mackerel *(Rastrelliger kanagurta*). Sardine (*Sardinella longiceps*), and non-fatty pelagic fish are suitable for Colombo curing. The fish is headed and gutted and washed of all dirt and slime. It is mixed with salt in the ratio of 3:1 or 4:1 (w/w) in large cement tanks. Pieces of a sour fruit, locally called "gorukapuli" (*Garcinia cambogia*), are mixed with the fish salt mixture. This gives acidity to the pickle. Processing time is 3-4 months. After this it is filled in wooden barrels along with liquid pickle up to the top and exported. Fruits of *G.cambogia* are yellow and contain tartaric acid. Smoke dried pulp, which is black, is used for pickling. The dried fruit is back, soft and acidic. It is extensively used for fish curry preparations in southern and western India. There are wide variations in the practice of the technique. In another well known processing practice, a dried pieces of gorukapuli is placed inside the belly flap of each fish. After this treatment, fish and salt are arranged in layers in large wooden vessels made of white cedar or mango wood. The vessels are filled up to the top, kept tightly closed upto 3 months, and thereafter sold in the open market. Colombo curing, although it became extinct in India, is still practiced in Sri Lanka where the process is known as ' "jadi".

viii. Mixed Salting

Salting process is low when the fish is large in size or is oily. As a result, it takes long for the self-brine to form and cover the fish. This may cause the fish in top layers to spoil before salt is picked up by them. An advantageous method of salting employed in such cases is mixed salting, also called pickle curing.

In mixed salting, particularly used for medium size fatty fish like herring, the fish is first mixed with dry salt and then packed in watertight containers with slat sprinkled between each layer. It is then topped with saturated brine. A cover is placed on top of the fish to hold them below the surface of the brine. Mixed salting has the advantage that the fish are completely surrounded by brine from the beginning, thus quickening the salting process.

2.2.3. Factors Affecting Salting Process and Quality of the Product

2.2.3.1. Source, Composition and Quality of Salt

i. Source

Pure common salt is sodium chloride. Commercial common salts may contain certain impurities, their levels depending on the source of salt and the method of production. The most common types of salt used are solar salt and rock salt.

- ☆ **Solar Salt:** Solar salt is produced by evaporating sea water or salt-lake water under the sun. Solar salt usually contains several impurities including a multiplicity of salts other than sodium chloride. Solar salt is quite often contaminated with sand and also may harbour high numbers halophilic bacteria.
- ☆ **Brine Evaporated Salts:** Underground deposits, which are brought to the surface in liquid form and evaporated by heating.
- ☆ **Rock Salt:** Rock salt is mined from natural deposits. This generally contains varying percentages of sodium chloride in the range 80-99 per cent. Rock salt is ground to varying degress of fineness before use.

ii. Chemical Composition

Suitability of the salt for fish curing will depend on its chemical composition, microbial purity and physical characteristics.

Commercial salts contain sodium chloride ranging, from 80-99 per cent. They may contain chlorides and sulphates of calcium and magnesium, Sulphates and carbonates of sodium, and traces of copper and iron and chemical impurities. Ca^{++} and Mg^{++} ions from impure salt bind with proteins and form a barrier to the Na^{+}

thus delaying the penetration of salt into the fish. An increase in the hydration of proteins and coagulation of gel occurs which are salted out of solutions. Magnesium chloride is hygroscopic and tends to absorb water. This will make it difficult to keep the fish dry.

Even a small quantity of 1 per cent Ca++ and Mg++ in salt imparts a white colour to the fish. However, even this small quantity imparts a bitter taste as well. Presence of 0.5 per cent of calcium and magnesium as sulphate in the salt can impart a desirable whiteness and rigidity to the fish flesh, but higher concentrations will lead to excessive bitterness and brittleness.

Salt used in fish curing should have at least 95 per cent sodium chloride. It should also be free from halophilic bacteria and as dry as possible. Calcium and magnesium impurities can be washed away using clean water as they dissolve quicker than sodium chloride and can be removed along with wash water. This should be resorted to only if purer common salt is not available.

iii. Microbial Quality

Most of the solar salts have been shown to contain a sizeable population of halophiles. One of the halophilic groups can cause a red or pink discoloration on wet or partially dried salted fish and hence is of great commercial significance. They do not grow is the fish are fully immersed in brine or fully dried. Halophilic moulds are met with more frequently in rock salt.

iv. Crystal Size

The size of salt crystals greatly influences the salting time. Coarse crystals dissolve only slowly and take more time to attain the desired only slowly and take more time to attain the desired concentration of brine. With large and coarse crystals the area of contact with fish will be less, adversely affecting its dissolution and salting. Too fine crystals may cause too much dehydration on the surface of contact with the fish leading to the consequent hardening and thus delaying further penetration of salt. This condition is called 'salt burn'. The size of salt crystals chosen shall be one that will ensure a faster rate of dissolution.

2.2.3.2. Nature of the Fish

Frozen and thawed fish allow greater permeability of salt into the flesh. The fresher the fish, the slower will be the diffusion of salt to the centre. Fish in rigor takes longer time for salt uptake compared to fish in which the rigor is resolved and is in the early stages of spoilage. Another factor determining salt uptake is the fat content. Higher amount of fat in the flesh acts as an effective barrier to salt uptake and slow down the process significantly.

i. Fish Size and Dressing

Salting time can be considered as reciprocal to its specific area. A flat fish with a large specific area will take less time for salting. Salting process can be greatly accelerated by splitting open the fish, thus providing greater specific are per unit weight for the salt to come into contact with. Still greater area of contact can be made available if deep scores are made in the split fish flesh. This will further accelerate the salting process. Thickness of the fish is another factor deciding the salt penetration. More the thickness of the fish, the slower will be the rate of diffusion of salt to the centre.

2.2.3.3. Salt Concentration in Brine

The rate of diffusion of salt into the fish flesh and that of water from the flesh tissues to the surroundings will depend upon the concentration gradient of salt in the two media. Greater the concentration of the salt in the surrounding brine, faster will be its uptake by the fish flesh. However, it need not always be strictly proportional to the concentration gradient, as different salt concentrations have been found to give rise to different changes in the protein ranging from hydration to salting out.

2.2.3.4.Temperature

Salt penetration into the fish flesh increases with increase in temperature. Increase in temperature intensifies the thermal motion of particles and also causes a reduction in the viscosity of water. But increase in temperature also may increase the rate of spoilage. If, however, the fish is salted at lower temperature, the rate of penetration will be reduced; but the rate of spoilage will be reduced even more.

2.2.4. Types of Salted Fish

Based in the quantity of salt used, salted fish can be divided into three categories – light salted, medium salted and, heavily salted.

Salting aims at saturating the water in the fish flesh wholly or partly with common salt. Salted fish can be saturated or unsaturated with salt depending on the quantity of salt used. The maximum concentration of salt attainable in the fish flesh is equivalent to that of saturated brine, *i.e.*, around 26 per cent. In practive, it will be lower because of the presence of other dissolved substances in the fish body water. Assuming that fish muscle contains approximately 80 per cent moisture, about 20-22 kg salt will be required 80 per cent moisture, about 20-22 kg salt will be required to saturate 100 kg of fish to get a heavily salted fish. The resultant product will contain 40-45 per cent salt on dry weight basis. If a smaller quantity of salt is used, the cure will be unsaturated.

For light salted fish 16-20 parts of salt per 100 parts of fish is used. The resultant product will contain around 20-30 per cent salt on dry weight basis. For medium salted fish 20-28 parts by weight of salt is used for every 100 parts fish and the resultant product will contain 30-40 per cent salt on dry weight basis.

In processing light salted fish, the duration of salting is of great significance. Too long a period of salting will lead to onset of spoilage of fish during the salting time itself. Drying or other types of further processing will have to be carried out immediately. Brown discoloration and rancidity die to fat oxidation and other putrefactive changes are quicker and more pronounced in light salted fish. Salting time generally does not exceed 24 hours for light salted fish.

2.2.5. Maturation in Salted Fish

Maturation is a phenomenon observed in brine pickled fatty fish during storage. During maturation the fish lose their raw flavour. The weight lost through loss of water by osmosis is regained through salt uptake within a few days and the fish flesh becomes tender and acquires a special pleasant taste. Maturing is considered to take place by the action of enzymes, proteolytic and lipolytic in that order. The enzymes responsible for maturation are derived from the digestive system of the fish, primarily from the pyloric caeca.

Maturation is also depended on the bacteria present in the fish as well as in the brine. Carbohydrates are fermented and provide the matured fish a pleasant flavour and slight sour taste.

The process of maturation is a highly complex one. The products of proteolysis and lipolysis are seen predominant in the product although several others like fermentation production of carbohydrates, the products of Maillard reaction *etc.* also contribute to the odour and flavour of matured fish.

2.2.6. Spoilage of Salted Fish

Bacteria exhibit different degrees of tolerance towards salt. Though salt is known to have bacteriostatic/bactericidal action, depending on its actual content many bacteria may become active. Most of the bacteria usually associated with fish spoilage are halophobic in nature and will not grow when the salt concentrationis in excess of 5 per cent. However, there are other organisms which are halophilic and can grow in environments of high salt concentrations. In addition to the action of halophiles changes occurring in the protein, fat *etc.* also will contribute to the spoilage of salted fish.

2.2.6.1. Microbial Spoilage

i. Pink' or 'Red'

This is a common type of spoilage associated with salted fish manifesting mostly during storage in warm weather conditions. The surface of the fish becomes covered with a red slime that gives off an unpleasant odour. This is brought about by halophiles which need salt concentration above 10 per cent for their growth. The spoilage is called 'pink' or 'red' because of the colour of the colonies of the bacteria appearing on the fish surface.

The microorganisms responsible for this phenomenon are halophilic rods and cocci originating from the salt, mainly solar salt used in the process. They include *Halobacdterium salinaria, H. Cutirubum, Sarcina marrhuae, S. Littoralis and Micrococcus rosens.* These are all aerobic organisms active only while in contact with air. They are thermophiles with an optimum temperature of growth of about 42°C. They will not grow at temperatures below 10°C.

The spoilage manifests first as a delicate pink sheen on the surface of the fish. At this stage it can be rubbed or washed off easily without damage to the fish. Fish will remain edible as the bacteria responsible do not produce toxin. However, if severely infected by the bacteria the red will reappear and the surface of the fish will become softened due to protein decomposition giving off ammoniacal odour. The flesh turns alkaline and become inedible.

The bacteria responsible being aerobic and thermophilic in nature, an ideal preventive measure will be to keep the fish out of contact with air and store at low temperatures, preferable below 10°C. After initial occurrence and washing off the slime, the recurrence of 'pink' can be prevented by treatment with formaldehyde or sulphur dioxide vapours or by dipping in a solution of sodium metabisulphite. Treatment with sodium/calcium salts or propionic acid also effectively controls the development of 'pink'.

ii. Moulds

Moulds are often seen to grown on salted and unsalted dried fish. Moulds will grow profusely if the RH is 75 per cent or more. They are also temperature sensitive having an optimum temperature for growth 30-35°C. 'Dun' is type of mould development observed even in heavily salted fish. This is characterised by the appearance of coloured spots, black, grew or brown, visible particularly on the fleshy side of the fish. This imparts an appearance as if sprinkled with ground black pepper. The small spots develop a root-like network into the interior of the fish flesh. This is caused by a group of moulds, of which *Sporendonema epizoum* is mostly proteolytic, but does not cause softening of tissues. However, the discoloration is unsightly and reduces the marketability of the product. *S. epizoium* is a true halophile and will flourish if the storage atmosphere us damp. It gas optimum growth at 10-15 per cent salt concentration, 75 per cent RH and 25°C. The mould activity itself may make the surface moist and pave the way for the fish becoming more susceptible to other types of spoilage.

The moulds can be relatively easily removed in the early stages. They can be easily scraped or brushed away; however they will reappear rapidly. In cases of severe infection dipping in 0.1 per cent sorbic acid will give some protection.

Solar salt is known as one of the sources of contamination; effective control may be achieved by use of good quality salt, maintenance of low temperature and humidity and, well ventilated and dry storage conditions.

iii. Saponification

The spoilage referred to as saponification is a damage caused by aerobic microorganisms active even at very low temperatures and is identified by a malodorous slime on the surface of the fish. This phenomenon is associated with light salted fish of the type, boxed herring, when in contact with air. In the initial stages of its development, the fish remains edible if washed with water and treated with a solution containing vinegar and water.

To prevent saponification the fish can be kept in brine containing vinegar for some time before boxing.

iv. Putrefactive Spoilage

If the salting process is very slow, it will take unduly long time for the salt to reach the centre of the fish and to saturate it. This can set in motion a putrefactive process in such regions. The flesh near the backbone become 'tanned' or reddened accompanied by the development of an offensive putrid smell. Tanning usually appears near the tail and kidney. Any presalting operation which can accelerate the penetration of salt to the interior of the flesh like gutting, splitting *etc.* can prevent the development of tanning.

2.2.6.2. Non-microbial Spoilage

i. Infestation by Maggots

Infestation by flies is a very common and serious problem faced by the salt fish trade. Infestation actually takes place during the early stages of drying. Cheese flies (*Drosophila casei*) are attracted to the drying premises due and the unhygienic atmosphere in the drying premises due to the presence of discarded and spoiled fish and other materials. Adult female flies lay their eggs on the fish and on the sides of the containers. The eggs hatch in a day or two and the larvae (maggots) jumps about and spread rapidly and soon infest the whole fish causing considerable damage to the fish. The larve metamorphose into red purpae from which fully-grown flies emerge. The emerging adults can then rein fest more fish and the cycle will continue.

Fish infested with maggots can be salvaged to a great extent if immediately washed thoroughly in brine and placed in clean uncontaminated containers. It is advisable to plunge the whole lot of fish into the brine when the maggots will float at the top. They can then be collected and destroyed. It fresh water is used for washing they will sink to the bottom and can be drawn away.

Infestation by cheese flies can be controlled in several ways. The best choice is minimising the chances of their coming into contact with the fish and fish handling premises. This include maintaining the handling and processing premises clean and fly-proof, ensuring that the fish is well covered during salting with no access for the flies to get in weather and multiplies freely at temperatures above 20°C, it is desirable to maintain the handling and processing premises cool as a control measure.

ii. Rust

Appearance of a colour similar to that of rusted iron on its surface is a phenomenon most commonly associated with salted fish. Oily fishes like sardine, mackerel *etc.* are particularly prone to rusting. Rusted fish is characterised by, besides its colour, an unpleasant taste and rancid odour. These are brought about by oxidation of oil in the fish by the atmospheric oxygen. The natural antioxidants in the tissues of live fish disintegrate soon after its death and hence the body fat becomes devoid of any protection. Further, the salt present in the salted fish accelerates the oxidation process. Fat is first oxidised to the hydroperoxides at which stage this cannot be detected organoleptically. However, the chain of reaction continues and the hydroperoxides break down to aldehydes, ketones, hydroxylic acids *etc.* These products are responsible for the rancid odours and flavours as well as the discoloration. Traces of iron and copper in the curing salt can catalyse the oxidation and consequent discoloration considerably.

Rust first appears on the surface and then penetrates the skin and spread throughout the flesh making it inedible. When it first appears on the surface, the impairment of the flavour is not significant. It can be washed away using dilute solution of sodium bicarbonate.

The best method to prevent the occurrence of rust in salted fish is to prevent it from having any contact with air. Fish should be kept covered with brine during salting and storage. Dry fish should be kept properly packed so as to avoid contact with air to the extent possible.

iii. White Spots

The presence of calcium and magnesium in the salt can produce remarkable whitening in the flesh of salted fish. Different from this, some white spots occur occasionally on the surface of salted fish which consist of crystals of disodium hydrogen phosphate derived probably from the enzymatic breakdown of nucleotides. The cause for the occurrence of white spots have been described as:

- partial drying of salt bulk prior to storage;
- low temperature of storage and
- initial spoilage of fish prior to salting.

Exposure to dry air will partially dehydrate the crystals leaving them as a scarcely noticeable white powder.

iv. Fragmentation

Cured and dried fish often become brittle and break during storage and transportation. This is referred to as fragmentation. Denaturation of proteins, hollowing of the fish by insect attacks, use of spoiled fish for processing *etc.* are the reasons ascribed to this. Fragmentation can be reduced by using fresh fish for processing and providing adequate protection against physical damage by using appropriate packaging.

2.3. Smoking

Smoking or smoke curing, like drying and salt curing, is an ancient method of preservation of fish. In the early methods of smoking, heavily salted fish used to be smoked for long durations, few weeks even, and the resultant products were called 'hard cures'. These products had long shelf life at ambient temperatures owing to the high salt concentrations and long smoking and drying periods which lower the water activity considerably. In course of time, the 'hard cures' gave way to milder products with less salt and lower durations of smoking. Such products, however, have only short shelf life. These are more relished because of their specific mild flavour and are considered delicacies. Smoking is also employed as an intermediate step in processing canned smoked fish. Lightly smoked fish is considered as an alternate to fresh having slight pleasant smoke flavour.

2.3.1. Fuel and Decomposition Products

2.3.1.1. Fuel

Fuel is the source of smoke as well as of heat. Smoking should impart, in addition to the specific aroma and taste, an attractive colour to the product. Fuel selected for generating smoking should be such that it meets the above requirements. Wood is the most commonly used fuel and therefore selection of the type of wood for fuel is a very important consideration in the smoking process. Wood is used in the form of sawdust, shavings or logs. It is also important that the wood does not impart any unpleasant flavour or colour to the product. Hard wood species are preferred as fuel for smoking. Soft wood species like conifers are not suitable because they contain volatile resins, which will impart undesirable flavour to the product.

2.3.1.2. Composition of Wood

Wood contains both combustible and non-combustible substances, the latter being constituted by water and ash. The combustible components are complex organic substances consisting mainly of polyoses (including pentosans) and lignin. Soft wood species will contain resins also in addition to the above. Combination of combustion and destructive distillation occurring while smoke is produced from smouldering wood results in a complex mixture of aliphatic and aromatic compounds in addition to water, carbondioxide and traces of hydrogen and carbonmonoxide.

2.3.1.3. Physical and Chemical Characteristics of Smoke

i. Physical Properties

On heating, wood undergoes destructive distillation and release gases and vapours in the temperature range 100-150°C. The vapour consists mainly of water. The volatile matter released in this temperature range constitutes only about 2 per cent of the total matter released. On raising the temperature to 200°C wood begins to char with an increased output of volatiles upto 25 per cent. Between 150 and 200°C the volatiles generated are mostly constituted by carbon monoxide and carbondioxide. Above 250°C the output of volatiles increases sharply and the quantum of hydrogen and hydrocarbons decreases. At about 300°C wood catches fire.

Smoke is a typical aerosol with the solid and liquid particles dispersed in a disperse gaseous medium. The gases produced during the process of destructive distillation condense in the cool zone above the fire to form a stable aerosol. The disperse medium is a mixture of gases like oxygen, hydrogen, nitrogen, carbon dioxide, carbon monoxide and various hydrocarbons. Many organic substances in the form of tarry droplets are present as vapour or as liquid depending on the source of smoke and conditions o smoke generation. The particle size, which also depends on the conditions of smoke generation and cooling, varies from 0.1 to 0.14 in radius.

ii. Chemical Composition of Smoke

☆ Products from Polyoses

Cellulose, the cellular fibre constituting the bulk of wood, is a polysaccharide $(C_6H_{10}O_5)_n$ having molecular weight approximately 150,000. At high temperatures it is hydrolysed to simple saccharoses, mainly glucose.

$(C_6H_{10}O_5)_n + nH_2O \quad n\,C_6H_{12}O_6$

Glucose undergoes further decomposition yielding oxymethyl furfural

$C_6H_{12}O_6 + 3H_2O$ OHC(O)C-CH_2OH, HC == CH

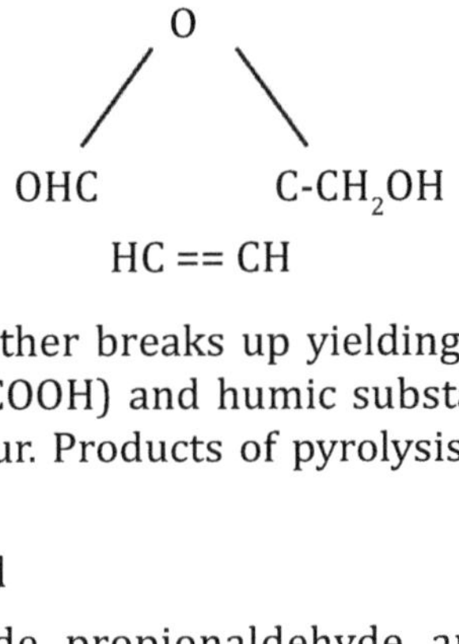

Oxymethyl furfural is unstable and further breaks up yielding formic acid (HCOOH), levulinic acid (CH_3-CO-CH_2-CH_2-COOH) and humic substances which give the smoked fish its characteristic colour. Products of pyrolysis of polyoses can be summarise as:

Alcohols: Methonal, ethanol, propanol

Aldehydes: Formaldehyde, acetaldehyde, propionaldehyde, arylaldehyde, furfuraldehyde, 5-methyl furfuraldehyde

Ketones: Acetone, diacetone

Acids: Formic acid, acetic acid, propionic acid

☆ Products from Lignin

Lignin is a constituent of the cell walls in the wood. Lignin is more resistant to heat and begins to break down at 350°C. Lignin contains methoxyl group CH_3O- and hence one of the products obtained on pyrolysis is methanol. Other products or pyrolysis of lignin are a complex mixture of phenols and tar. The phenolic constituents consist mainly of guaiacol and its 4-methyl and propyl homologues, catechol and its 4-methyl derivative, pyrogallol and its methyl, dimethyl ethers and homologues and, hydroquinone, phenol and p. cresol.

iii.Tar Distillate

Tar distillate contains both low and high-density tar oils. Low density tar oils having boiling points below 140°C contains valeric aldehyde and furanes – furane, methyl furane, dimethyl furane and trimethyl furane. Tar contains guaiacol, vinyl guaiacol, cresol, catechol, phenol, hydroquinones, eugenol and similar compounds.

High density tar oils have boiling points above 200°C and contain phenols and their derivatives like 0-*, m- and p- cresols, xylenols, ethyl phenol, catechol, guaiacol, esters of pyrogallol and lignoceric acid. The aromatic substances in the tar are products of lignin decomposition.

2.3.1.4. Products of Complete Combustion

When combustion is incomplete, smoke produced contains organic substances as above which can impart a distinct smoky flavour to the fish. But when the combustion is complete organic products break down yielding carbon dioxide and water with no formation of smoke. Therefore when the fuel, which is often saw dust, is burned, the temperature and supply of air are to be so controlled that the fuel burns incompletely and adequate smoke is produced.

2.3.1.5. Benzopyrene in Smoke

Wood smoke may contain traces of a polynuclear aromatic hydrocarbon 3,4 benzopyrene (CH_{10}) which is a carcinogen that can cause tumours in humans and animals. Its quantum may vary depending upon the method of smoke generation. Stronger draught and high combustion temperature are two factors contributing to its increase. However, no harmful effect has been experienced due to this, presumably due to its extremely low concentration. It is also probable that benzopyrence, if present, would tend to concentrate in the particle phase and would not be absorbed by the fish.

2.3.2. Deposition of Smoke

As the disperse medium of smoke is not very viscous, intense Brownian movement of the dispersed solid and liquid particles occurs. The colloidal particles collide, coagulate and form flakes. Smoke is stabilised by polar compounds like phenols, amyl alcohol *etc.* in the surface vapours. With maximum stabilisation the adsorption layer is completely saturated. Further concentration of stabiliser results in partial coagulation of aerosol. Water is a polar substance and with high relative humidity the smoke will coagulate with tarry substances setting on the surface of the fish.

During smoking largest concentration of smoke is seen at the top of the kiln. Accordingly, deposition of tarry substances will be more on fish at the top than on those in the lower layers.

2.3.3. Preservative Action of Smoke

Preservation by smoking is due to the combined action of salting, drying and deposition of natural chemicals produced during the thermal breakdown of wood. The roles played by individual constituents of smoke are not fully understood, still the importance of several components can be explained with reasonable accuracy.

- ☆ Salting which is an essential step prior to smoking proper, reduces the a_w and inhibits the growth of many spoilage microorganisms as well as pathogens
- ☆ Smoking causes partial surface drying of the fish. This will present a physical barrier to the passage of the microorganisms and also ratard theproliferation of aerobic microflora. Smoking at higher temperatures, as in hot smoking, wil destroy the enzymes as well as bacteria.
- ☆ Phenolic constituents like guaiacol and its homologues are effective antioxidants providing protection against autoxidation of lipids.
- ☆ Antimicrobial agents like phenols, formaldehyde *etc.* deposited on the fish control the microbial spoilage to a great extent and are considered responsible for the biological stability of smoked fish.

2.3.3.1. Bactericidal Property of Smoke

The bactericidal properties of smoke is due to the presence of chemicals such as formaldehyde, phenols, organic acids and other organic compounds that make up the tarry constituents of the smoke. Smoke behaves differently towards pure cultures of microorganisms and microflora of fish. Pure cultures of non-sporers like streptococcus are killed by smoke in three hours, whereas those of spore-formers like *Bacillus subtilis* and *B. mesentricus* survive even beyond seven hours exposure to smoke. Wood tar phenols are most active inhibitors of *Staphylococcus aureus*. As regards the effect of smoke on the microflora of smoked fish it is known that smoke below 40°C has only a weak antibacterial action. It is only during hot smoking that the number of microorganisms is considerable reduced, which is mainly due to the high temperature of smoke.

2.3.3.2. Significance of Phenols

i. Colour Formation

One of the indices for deciding the adequacy of smoking fish is its phenol content. The formation of colour in smoked fish is considered related to the deposition and subsequent oxidation of phenols. Oxidation of phenols is accelerated under alkaline conditions.

ii. Antioxidant Effect of Phenols

Wood smoke is also known to have a pronounced antioxidant effect providing effective protection against autoxidation of fat in fish. This is attributed to guaiacol and its homologues. 1, 3 dimethyl pyrogallcol also contributes to the antioxidants catechol, 4 methyl catechol and pryogallol 1-0- methyl ether from the particle phase of electrostatic smoking provide greater protection against autoxidation.

2.3.4. Generation of Smoke

2.3.4.1. Smoke from Burning Wood

Wood is the source of smoke used the world over. Wood in the form of chips or coarse shavings or as saw dust is most commonly used. Wood chips and sawdust is an ideal combination. A layer of wood chips or coarse wood shavings with a layer of sawdust atop is used for generation of smoke. As air cannot get easy access gto fire, the sawdust smoulders rather than burns. Lower temperature and limited supply of oxygen causes production of smoke with high content of flavouring and preserving principles.

2.3.4.2. Frictional Smoke Generator

Another method of production of smoke is using frictional smoke generators. In this process the end of a piece of wood is pressed against a rapidly revolving drum of dist. Due to friction the temperature of wood rises up to 350°C and smoke is generated. Frictional smoke is generate at a temperature lower than that of charring smoke. Frictional smoke contains approximately four times the quantity of volatile acids, eight times the quantity of volatile aldehydes and ketones and three times the quantity of phenols compared to smoke from charring wood.

2.3.5. Preparation of Fish for Smoking

2.3.5.1. Splitting, Gutting

Quality of the raw fish the significant influence on the quality of the smoked product. Pelagic fishes are considered best for smoking when their fat content is the highest; however, depending on the end product desired, fish of different fat contents are preferred. Generally fish for smoking should be kept chilled. In certain cases of non-fatty fishes it is considered desirable to keep the fish iced for two day before smoking.

Fish may be smoked whole or after splitting, gutting *etc.*, the methods employed being dependent on the end product desired, *e.g.*, smoked herring is processed in split and ungutted form. In some products the fish is split with the backbone retained on one fillet. Large fish are gutted, headed and split or made into fillets. It is very important that the guts, gills, kidney *etc.* are completely removed. These are the areas where spoilage can take place quickly and can affect the overall quality of the end product, particularly the light smoked fish which are more perishable because of the low salt content and light smoking at low temperature.

2.3.5.2. Salting

Fish is salted either by mixing with solid salt or by immersing in strong brine. Size of the fish, fat content, whether the fish is whole, gutted, split or cut, presence or absence of skin and other requirements in the end product *etc.* are the factors considered in determining the duration and type of salting. Depending upon the concentration of brine there may be an uptake or loss or water in the fish. However, there will be no appreciable difference if the saturation of the brine used is 70-80 per cent and for this reason, this is the most common concentration of brine employed. Fish salted in 100 per cent saturated brine is not preferred as it is likely to make the product unattractive because of the powdery, unattractive, crystals of salt that may be left on the surface. In weaker brines longer periods of immersion are needed and there will be an uptake of water which will have to be evaporated during drying. Salting is so adjusted that the fish takes up 2-3 per cent salt which is the optimum requirement in many smoked fish. However, when the fish is dry salted the uptake of salt will be very high. Such fish are soaked for several hours in water before smoking to bring down the salt content to the desired level.

2.3.5.3. Hanging

After salting the flesh of the fish will be moist. Surface proteins will dissolve in the brine yielding a sticky solution. For efficient smoking fish must be dry enough with no free water on the surface. Free water can cause condensation of the distillation products, particularly tar, which leave dark brown colour in the product. During hanging the sticky solution dries on the surface and makes the skin glossy. However, excessive drying will prevent the fish from open air or a hot air drier or even directly in the smoke chamber by burning wood without producing smoke.

Split and salted fish should remain in a stretched out position for uniform exposure of all parts to smoke. Hanging is done in several styles depending on the fish handled and the end product desired. Fish like sardine meant for canning after smoking are threaded through the eyes on thin speats. Large split whole fish like salmon are suspended by the tail and kept flat open by means of sticks or skewers threaded through the flesh. Fish can be hung or even spread on wire mesh trays, however, care has to be taken to ensure that the mesh does not leave excessive mark on the skin or flesh.

2.3.6. Smoking Process

Two types of smoking are popular, cold smoking and hot smoking. The essential difference between the two processes is the amount of heat the fish receives during the process. Cold smoking is a mild process where temperature is not raised enough for the fish flesh to undergo even a partial cooking. In hot smoking the temperature of smoke is high and the fish flesh is cooked even to the extent of partial sterilisation. However, this partial sterilisation does not result in any increased shelf life as the fish may get reinfected with microorganisms during subsequent handling and packaging.

2.3.6.1. Cold Smoking

Cold smoking is still carried out in more or less the conventional style, mostly using traditional chimney kilns. Fish is kept hung or spread in mesh trays in an upward draft of smoke produced in the floor by the burning fire. In the initial stages the fish is moist and the smoke is highly humid. Under these conditions use of high temperature will the invariable result in cooking of the fish flesh. To avoid this, the temperature inside the chamber should not be raised to the maximum employed in the process. In the second stage, when the surface of the fish is considerably dry, the temperature inside the chamber could be raised to a level that the fish species concerned will tolerate. However, it is desirable to complete the process at the same temperature throughout. The highest output of good quality products is quantities of ventilated smoke are maintained in the correct proportion.

During smoking the fish dries as also absorbs the aromatic substances from the smoke. The relative humidity inside the chamber also is decisive factor in determining the quality and output of the products. RH above 70-80 per cent will considerable slow down the drying and smoking process. To control the RH of the smoke, the flame dampers may be kept open thus providing a good draught whenever the per cent RH goes beyond the required level. Cold smoking is generally carried out at temperature not exceeding 40°C. Duration of smoking extends to several hours, say 36-72.

2.3.6.2. Hot Smoking

Several designs of kilns or smoking tunnels are available for hot smoking fish. The fuel is burnt either directly inside the kiln on movable trolleys or in external hearths located near the tunnel. The fish is charged into the tunnel in cages. The chamber will be having a metal frame structure and brick walls and can hold a number of fish cages.

The 'Torry kiln' designed and developed by the Torry Research Station, Aberdeen, is a very popular and is widely used in the UK for fish smoking. This is a horizontal flow kiln and such kilns are often called mechanical kilns because an electric fan draws the smoke over the fish in racks. Electric heaters are placed at different points in the smoke path to maintain even temperature throughout the smoke chamber.

In contrast to cold smoking where cooking of the flesh is neither the aim nor is attained in any case, in hot smoking fish is dried and cooked in the kiln before it is smoked. Drying is done in an intense draught of hot air at 75 to 80°C produced by burning fire. The skin of the fish becomes dry while the flesh becomes cooked as well.

At this stage the fish is considered ready for smoking. Smoking is produced by covering the burning logs with sawdust and the temperature in the chamber is maintained at or above 100°C. A schedule of operation in hot smoking can be considered as drying at 75 to 80°C for an hour followed by smoking at 95 to 100°C for an hour followed by smoking at 95 to 100°C for another one hour. The requirements will vary with the size and species of fish.

2.3.6.3. Electrostatic Smoking

A process of smoking developed and popular in the Soviet Union is electrostatic smoking. Smoking takes place as a result of the electrokinetic properties of smoke in a high voltage field of the order of 40 kW or more. Salted and rinsed fish is passed through a drying chamber heated by infrared lamps positioned on either side of the belt carrying fish. Fish is heated for 3 to 4 minutes at 40 to 50°C in the chamber when they lose around five per cent of its weight. They are then taken to the electrostatic smoking oven over a conveyor. On either side of the conveyor are Nichrome electrodes suitably spaced. The bases of the electrodes are fitted with high capacity heating elements to prevent condensation of moisture. The electrodes are supplied with high voltage current at 30 to 70 kW while smoke is admitted to the bottom of the oven. For an appealing smoked appearance of the fish at least one mg of smoke substances must be precipitated per square centimetre of fish surface. This requirement is attained in less than 5 minutes in the oven.

From the smoking chamber the fish are carried to the baking oven fitted with sufficient number of emitters of required capacity. In 5 to 6 minutes in the baking oven the fish are heated to about 30°C and lose 10 to 12 per cent of moisture. The advantages of the process include

- ☆ considerable saving in time as the whole process takes only around 20 minutes
- ☆ consequent increased output
- ☆ reduction the losses because of the short processing time
- ☆ The process is continuous and carried out in a mechanical equipments

2.3.6.4. Use of Smoke Concentrates

Smoking fluids or smoke concentrates' derived from dry distillation of wood are sometimes used to impart smoked colour and flavour to fish. The salted fish after washing and passing under an air blast is given a short dip of few seconds in the smoking fluid. The fish is then dried in mixture of smoke and air. During drying the fluid thickens and polymerises coating the fish surface with a thin film with typical golden yellow colour. The process cut shorts the smoking time, increases the productivity and results in considerable saving in materials and outlay. Improved antioxidant effect is considered another advantage. However, the product is rated organoleptically inferior to fish smoked in the normal way.

2.3.6.5. Use of Artificial Colouring

It is often difficult to achieve uniformity in the product, particularly colour, in smoked fish. Even if near uniformity in colour can be achieved, batch to batch variation occurs. One way of overcoming this inherent deficiency of smoking process is to artificially colour the fish with permitted dyestuffs. Artificial colouring is done by incorporating suitable dyestuffs in the brine used for salting. Vegetable dyes, coal tar dyes *etc.* are so used, the selection being dependent on the type of the product.

2.3.6.6. Post-Process Handling

On removing from the kiln, the smoked fish is warm and is generally allowed to cool before grading and packing. Cold smoked fish may lose upto 5 per cent of its weight by further evaporation. If the smoked fish is packed while warm, moisture will condense and deposit inside the containers. This will create a situation conductive to premature mould growth. Smoked fish can also be frozen and cold stored which will minimise the chances of spoilage before consumption.

2.3.7. Smoked Products

i. Masmin

The maximum priced smoked fish in India is masmin. Masmin is prepared from tuna. Other than this smoked sardine flake prepared from sardine is also gaining importance.

Masmin is a traditional smoke cured product of Lakshadweep Islands and Maldives. Lakshadweep has flourishing tuna fishery. Tuna is preferred for the preparation of masmin as it has large amount of myoglobin in the fish.

Preparation

Fresh tuna (skipjack) immediately after landing is beheaded, gutted and washed thoroughly. The fish is then boiled in 10-12 per cent salt water for one hour. The entire fish is immersed in salted water and the fish should be boiled until the red colour of the meat changes into light pink colour. The bones and skins are then removed and the fish is made into four chunks. The chunks are arranged in the smoking trays and kept in the smoking kiln. It should be smoked for two to three hours until the colour of the fish becomes golden brown smoke has antioxidative and bactericidal properties and compounds impart pleasing colour and flavour. After smoking, it is dried under sun to reach a consistency of wood. It is stored in polythene bags or polythene woven gunny bags and marketed.

Table 2.5: Characteristics of Masmin

Method	*Moisture (Per cent)*	*Protein (Per cent)*	*Ash (Per cent)*	*Fat (Per cent)*	*Sodium Chloride (Moisture free basis) (per cent)*
Traditional	6.8	66.30	8.1	19.0	7.6
Improved	10.2	76.60	9.5	5.7	1.6

iii. Katsuo-Bushi/Katsuo Fush (Dried Bonito Sticks)

Fish "fush" are a special product in japan. they are prepared by boiling, broiling (smoking) and drying fish fillets.

The procedures for preparation of bonito "fushi" are as follows: (a) cutting of raw fish (heading,filleting and splitting), (b) placing the cut fish meat (strip) on a tray, (c) boiling, (d) cooling and removal of small bones,(e) first broiling (smoking),(f) pasting cracks, (g) second broiling (smoking), (h) first drying in the sun, (i) shelving, (j) second drying in open air, (k) allowed to develop mold several times. about three months are required to complete the processing.

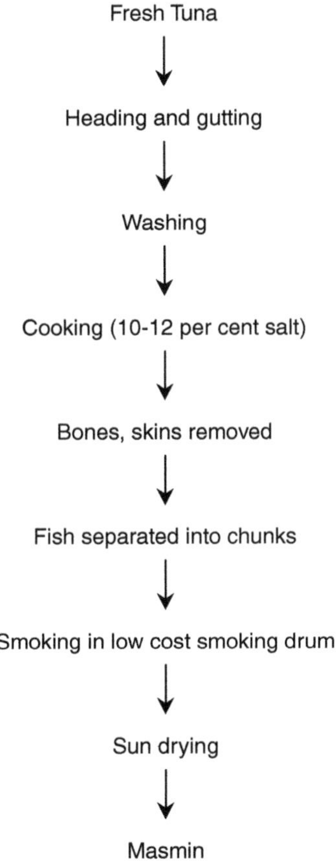

Figure 2.9. Flow Chart for the Preparation of Masmin.

2.4. Marinades

Marinades are, by general definition, fish or fish portions processed by treatment with edible acids and salt, and put up in brine, sauce or oil. Adjuvants for seasoning and bleaching are also often used. Fish marinades are characterised by the typical marinated odour and flavour. Treatment with acetic acid and salt brings about a sort of refinement in the fish simultaneously developing the typical marinated odour and flavour. Further improvement in odour and flavour is brought about by addition of various spices and covering liquids. Pelagic fatty fish such as sardine, herring *etc.* are generally used to process marinades, though fish and shellfish. In order that the characteristic odout, flavour and taste of marinades are maintained, use of additives which can suppress spoilage and prolong shelf life is restricted. Therefore, marinades have only a limited shelf life and are classified among the semi-conserves.

2.4.1. Preservative Action of Acetic Acid and Salt

2.4.1.1. Action of Acetic Acid

Development of putrefactive microorganisms is greatly delayed in an acetic acid medium having acid concentration of 1-2 per cent. If the concentration of acid is higher, many putrefactive bacteria will perish. It is only some microorganisms like moulds and yeast that will thrive in an acid medium. Moulds can gradually decompose acetic acid thereby creating a favourable atmosphere for the development of putrefactive bacteria. On the positive side, the characteristic succulence and tenderness of the marinated fish flesh is produced by the action of acetic acid which favours the action of some of the proteolytic ezymes. The action of these enzymes is responsible for the partial breakdown of proteins and release of some free amino acids which impart that characteristic flavour and taste to the marinated products. Acetic acid in the concentrations employed lower the pH of the meat to the required level creating an atmosphere for the autolytic reaction to take place. At higher concentrations, these reactions will be arrested and hence only lower concentration of 1-2 per cent acid is used.

2.4.1.2. Action of Salt

Salt also has different functions to perform in a marinade. Salt extracts water out of the fish tissues and causes the proteins to coagulate allowing the proteolysis to proceed at the desired level. Further, salt and acetic acid act in a mutually complementary manner. For example, yeast develops very rapidly in an acetic acid medium; however, stops growing at a 14 per cent salt concentration when the pH is maintained at 2.5. An equivalent effect is obtainable in a natural medium only when the salt concentration is 20 per cent or above. Thus the combined action of acetic acid and salt can favour employment of lower concentrations of both.

2.4.2. Types of Marinades

On the basis of the process employed the marinades are classified into three groups:

- ✰ Cold marinades
- ✰ Cooked marinades
- ✰ Fried marinades

2.4.2.1. Cold Marinades

As the name indicates, cold marinades are a class of products processed in the cold. The fish and other ingredients are not subjected to any heat treatment at any stage in the processing schedule. Both processing and storage are carried out at temperatures less than the ambient, say 10-12°C.

Cold marinades are the marinades proper and are processed mostly out of fish like herring, sprat and sardine.

Preparation of Cold Marinades

The raw fish is first washed in a solution containing 8-10 per cent salt for about an hour to remove the slime and to harden the meat. It is then headed, gutted, filleted or deboned as required. The fish or fish portions are again placed in a bath containing 3-5 per cent salt for about 30 minutes. This will facilitate removal of traces of blood from the muscle. A small degree of bleaching desirable in the product can be achieved by adding 0.5 per cent acetic acid to the salt bath. Following this pre-treatment the product is introduced into the finishing bath which is also called the softening bath.

The finishing bath consists of a solution of salt and acetic acid, their strength being determined by the ratio of the fish to solution and the type of the end product desired. Two important chemical processes take place in the fish in the finishing process. One is the reaction of acetic acid producing the characteristic tenderness in the marinade which is largely due to the proteolysis induced by autolytic tissue enzymes, and the butter-like taste by the liberation of amino acids. The other is the action of salt causing removal of water and coagulation of proteins. The salt also inhibits hydrolysis and causes it to proceed within desired limits. The development of aroma is ascribed to microbial action. Therefore the ratio of fish to pickle is so adjusted that at the end of the process the finishing bath has an acetic acid concentration of at least 2.5 per cent. The general practice is to use salt 1-7 per cent higher than that of acetic acid depending of the strength of the final bath. A usual strength of the solution used is 7 per cent acetic acid and 14 per cent salt. It the fish is held at chill temperatures, the strength can be less, however, the concentration must be increased during warmer seasons of the year to prevent spoilage. If the acid and salt levels are too high, the microorganisms cannot survive and the characteristic flavours may not develop.

Pickling is done at an optimum temperature of 10-12°C. Higher temperature will favour the process proceed more rapidly, but the flesh will become too soft and its succulence will lost. At lower temperatures, development of the characteristic flavours is retarded. Pickling is generally concluded in 3-7 days. Pickling is carried out in wooden vats or barrels, open or closed. In any case the container should be full and have a tight fitting lid to prevent contact with air and the consequent development of rancid flavours.

The fish at the end of the process should be firm, white, opaque and tender. It should not be glassy or shining in the innermost portion of the flesh and should be free from any red spots. The marinated fish are packed often in glass jars, and covered with a solution containing 1-2 per cent acetic acid and 2-4 per cent salt.

The acid taste of the cold marinades may be reduced by using citric or tartaric acid replacing some or all of the acetic acid. However, the pH of the final solution should be maintained at 4.5. To improve the flavour and soften the acetic acid taste, the acetic acid-salt solution is often flavoured with spices like pepper, clove, coriander, bay leaves *etc.* Use of spice extracts is considered better as they, in general, will be free of bacteria. Slices of onion, carrot, celery *etc.* may serve as garnish.

Shelf life of cold marinades may be several months at chilled temperatures, but in tropical conditions it may be only a few weeks.

2.4.2.2. Cooked Marinades

Cooked marinades are generally packed in a jelly and are also called jellied products. The fish or fish pieces are treated with edible acids and salt and are heat treated. The combined effect of heat and acid in the bleaching bath is considered the preservative principle in the cooked marinades. Low pH is often maintained in cooked marinades to prevent the growth of harmful bacteria like *Clostridium botulinum*. The shelf life of cooked marinades can be upto six months.

Preparation of Cooked Marinades

The fish is dressed and salted as for cold marinades. After draining off the brine the fish is spread on perforated trays and immersed in a bleaching bath containing 1-2per cent acetic acid and 6-8 per cent salt maintained at 85°C for 10-15 minutes. During this process fish suffers a weight loss of about 15-20 per cent in the form of water removed from it. This will dilute the bleaching bath and hence it must be renewed from time to time. At the end of immersion the trays are taken out of the acid bath and rinsed free of adhering fat and protein foam. Product is packed in appropriate containers made of glass, porcelain or lacquered cans. The concentration of acetic acid in the jellied brine should be about two percent, and the salt content about three percent. Generally a 4-5 per cent gelatin is considered appropriate in the jellied brine. In preparing jellied brine the temperature should not exceed 70°C as gelatin is very heat sensitive. Spices and saccharin can be added to the jellied brine. The brine is cooled to 4°C before filling.

2.4.2.3. Fried Marinades

In the fried marinades the fish after pre-treatment is baked or broiled in oil with or without breading and is covered with acetic acid or sauces. The higher temperature of frying kills most of the bacteria in the fish flesh. If stored at 0-8°C fried marinades can have a storage life of upto one year. Fried marinades are made out of different fish.

Preparation of Fried Marinades

The fish is dressed and salted as for cold marinade; but usually with the backbone retained. After the fish or fish pieces are drained well, they are treated twice in a breading mixture of equal parts of wheat and dry flour. Leaving for an hour after first breading the fish is again given a second coat of breading. Any excess flour is gently tapped off.

The breaded fish is fried for 5-12 minutes in oil maintained at 160-180°C. The frying oil consists of a mixture of oil and solid fat to obtain a pliable film of fat all around the fish. The duration of frying depends on the temperature of frying oil, thickness of fish and its water content. In a deep fat frying system the termination of the frying process is indicated by the rise of the fish to the surface, because the specific gravity of the fish is altered by the absorption of fat and loss of water. During frying the interior of the fish flesh attain temperature below the boiling point of water, though higher temperature persists in the oil medium. The breaded outer layer attains high temperature, and the flesh takes on a characteristic dark colour and special aromatic components of fried products are formed.

The fried fish is cooled well and packed in cans. It is then covered with brine containing 3-5 per cent salt and 2-3.5 per cent acetic acid. Use of malt vinegar or vinegar containing spices and herbs is recommended for fried marinades.

The ratio of fish to covering liquid is about 2:1. Much of the water lost by the fish during frying will be taken up from the covering liquid which will take place gradually. Therefore, the filled cans are allowed to stand open for several hours and are then again filled to the brim and then sealed.

2.4.3. Shelf Life of Marinated Fish

Marinades in general are stable only for a limited period. The only preservatives in cold marinades are acetic acid and salt. In cooked and fried marinades, even though there is a heat treatment involved, the temperature attained by the flesh is below 100°C which is not sufficient to kill all microorganisms. There is also the possibility of post-process contamination up to the time the cans are sealed. The shelf stability depends mainly on the storage temperature and, to some extent, on the type of bacteria present. Spoilage in cold marinades is indicated by gas formation; gas formation and liquefaction in jelled products, and gas formation with appearance of slime in fried marinades.

Spoilage occurring in fish marinades is classified into physical, chemical and biological types.

2.4.3.1. Physical Spoilage

If the pack of marinades is stored below 0°C the contents will freeze. Water expands on freezing and may damage the glass jar or tin can. In Cooked marinades the structure of jelly will be damaged by the formation of ice crystals. Heat damage will occur if the cans are filled when the product is too cold, and stored in warm environment. Cans may swell if cooked marinades are sealed in cans before swelling of the flesh is completed. However, the contents of the can are very rarely altered by physical spoilage.

2.4.3.2. Chemical Spoilage

Chemical Spoilage is caused by the action of acetic acid on the metal of the badly lacquered and/or tinned cans. Because of the corrosive action of the acid on the metal, hydrogen gas will be liberated which will cause the cans to swell. Metal dissolved in the product will also impart a metallic taste to the product and alter its flavour making the product unacceptable.

2.4.3.3. Biological Spoilage

Biological spoilage may result in simple flipping to hard swells and even in bursting of the cans. Biological spoilage in fish marinades is of two categories, one due to the microbial degradation of proteins in the fish and the other due to the fermentation of the added sugar or other sugar containing additives. Biological spoilage also results in production of gas, mainly carbondioxide and also in the degradation of the quality of the product. Distinction between chemical and biological spoilage can be made by analysis of gas, bacteriological and chemical investigations and also by organoleptic evaluation.

The principal causes of spoilage in marinated fish are heterofermentative lactic acid bacteria in cold marinades, proteolytic bacteria in cooked marinades and slime formers in fried marinades. High content of acetic acid and salt can control the action of putrefactive bacteria and slime formers. Suppression of lactic acid bacteria is very difficult, as it will require very high concentrations of acetic acid. Fish marinades are distinguished for their taste and aroma and conditions required to arrest the bacterial action, that is use of high concentrations of acid and salt, will impair both taste and arome. Therefore the most important requisite in processing marinades is high grade raw materials supplemented with maintenance of strict hygiene and sanitation in the premises and processing schedules including work table, utensils *etc.* throughtout the process.

2.4.4. Other Types of Marinated Fish

There are certain other products, which are considered as marinades though they do not strictly come under this group of products. The South American specialties called Ceviche and Escabeche, and Paksiw of the Philippines, are Prominent among them.

Ceviche is made by marinating fish or shellfish with sour orange juice, lime juice or tartaric acid dissolved in water. To process ceviche skinless fillets of fish is diced into 10-15 cm cubes, combined with a mixture of onion and garlic, and seasoned with salt. The marinating solution, *viz.* the orange/lime juice or tartaric acid solution, is poured over this and let to marinade over night. The product has a shelf life of less than five days in tropical temperatures. For making escabeche, fish pieces are fried and marinated in vinegar. Vinegar may, often, be flavoured with garlic, ginger and pepper. Sugar also may be added.

For paksiw, raw fish together with coconut, vinegar, salt and sometimes sugar are boiled together in a pot. Water is added and the mixture is re-boiled. The product has a shelf life of several days.

2.5. Hurdle Technology in Fish Preservation and Processing

Hurdle technology (also called combined methods, combined processes, combination techniques or barrier technology) advocates the deliberate combination of existing and novel preservation techniques in order to establish a series of preservative factors (hurdles) that any microorganisms should not be able to overcome. Spoilage and poisoning of fish and fishery products by microorganisms is a problem that is not yet under adequate control, despite the range of preservation techniques available (*e.g.* freezing, blanching, pasteurizing and canning). In fact, the current consumer demand for more natural and fresh like foods, which urges food manufacturers to use only mild preservation techniques (*e.g.* refrigeration, modified-atmosphere packaging and biopreservation), make the preservation even greater difficult task.

2.5.1. Hurdles Employed in Fish Preservation

Upto now about 50 hurdles have been identified for use in food preservation. Commonly used hurdles are high temperature, low temperature, low water activity, acidicity, low redox potential, competitive microorganisms and preservatives. Emerging hurdles are ultra high pressure, MAP, bacteriocins, edible coating.

2.5.2. Potential Hurdles in Fish Preservation

Physical Hurdles

- ☆ High temperature (sterilization, pasteurization and blanching)
- ☆ Low temperature (Chilling and Freezing)
- ☆ Ultra violet radiation
- ☆ Ionizing radiation
- ☆ Electromagnetic energy (Microwave energy, Radio frequency energy, Oscillating magnet field pulses and high electric field pulses).
- ☆ Photodynamic inactivation
- ☆ Ultra high pressure
- ☆ Ultrasonication
- ☆ Packaging film (Plastic multi layer, active coating and edible coating),
- ☆ Modified atmospheric packaging (Gas packaging, Vacuum packaging, moderate vacuum and active packaging),
- ☆ Aseptic packaging,
- ☆ Food microstructure

Physical Chemical Hurdle in Fish Preservation

Low water activity (a_w), low ph, low redox potential (eh), salt, nitrate, nitrite, carbondioxide, oxygen, ozone, organic acids, lactic acid, lactate, acetic acid, acetate, ascorbic acid sulphate, smoking, phosphate, glucono icotones, phenols chelators, surface treatment agents, ethanol, propyl glycol, maillard reaction products, spices, herbs, lactoperoxidase and lysozyme.

Microbiology Deried Hurdles

- ☆ Competitive flora
- ☆ Protective cultures

- Bacteriocins
- Antibiotics

Miscellaneous Hurdles

- Monolaurin
- Free fatty acids
- Chitosan
- Chlori

2.6. Fermented Products

Fermentation is an age-old technology of preservation of highly perishable freshwater and marine animals. However, processing of fermented fish and fishery products is almost exclusively confined to the Asian region. Only very few of these products are known to be processed outside Asia.

Most, if not all, processes employed in fish preservation are so designed that the identity of the fish is unaltered, and aim at keeping the fish meat as near as possible to its original state. If the fish is adequately protected against contamination by microorganisms and endogenic enzymes, the flesh will not break down and remain remarkably stable for a considerable period. However, fermentation involves breakdown of proteins in the raw fish to simpler substances storage. Fermentation is generally described as a process in which the complex protein molecules in the fish are broken down by the action of organic catalysts, enzymes or ferments into simpler molecules. In some processes where the process is controlled by adding salt, only a partial breakdown of the proteins takes place. Such processes are so controlled that a product with a desired type of flavour is produced simultaneously ensuring preservation of the product. Breakdown of protein can be brought about by the action of exogenic as well as endogenic enzymes, the latter being present in the guts and intestines of fish. Sometimes microorganisms are also involved. Fermented fish products have a meaty flavour and provide dietary needs of amino acids, some vitamins and other nutrients.

2.6.1. Fermented Fish Products

The original fermented fish product was perhaps, the liquid exuded from fish during salting. Fermentation of fish without addition of salt, but employing other agents, is also popular. Different processes employed in the fermentation of fish yield three distinct types of products as follows:

- products in which the fish substantially retains its original form
- products in which the original fish is reduced to the form of a paste and
- fish sauces in which the fish meat is reduced to a liquid.

Fermented fishery products are generally classified into three categories, high-salt (20 per cent or more salt), low-slat (6-18 per cent salt) and no-salt products.

High-salt (20 per cent or more) -	Liquid separated fish sauce
	Residue cured fish
	Mincing and partial drying fish paste
Low-salt (6-18 per cent):	Lactic acid fermentation
	Acid pickling at low temperature
No-salt:	Dried bonito fermentation
	Alkaline fermentation

Though most mammalian enzymes are highly proteolytic and active at low pH, the visceral and digestive tract enzymes in the fish are active at or near neutral pH. Various cereals and plants are added in many cases so that the digestive enzymes from these sources also aid in the process of fermentation.

Fish subjected to salting alone is likely to ferment to some degree. The degree of fermentation will depend on the proportion of salt used, fat content of the fish, dressing of the fish like complete or partial removal of the gut, the nature of the additives, if any, added and the temperature at which the salted fish is maintained. The temperature is of particular importance. This has been evidenced by the observation that using precisely the same process, but by maintaining at different temperatures, products quite different in characteristics can be obtained.

With salt concentration higher than 20 per cent three different types of fermented products can be obtained. If fermentation is allowed to complete, *i.e.*, when all the protein matter has been broken down to simpler water soluble compounds, the resultant liquid is called fish sauce. When it is partial, the fish is likely to substantially retain the original form in which it was treated with salt. The resultant product is cured fish. This product is minced and partially dried or otherwise processed to yield fish paste.

When the concentration of salt is less than 20 per cent as in the case of low-salt process, the fish may undergo spoilage due to the action of micoorganisms. Therefore, some means of preservation other than the use of salt is also essential. In a low-salt process, lactic acid fermentation along with added sources of carbohydrates is a common practice. Rice, milk, flour and syrup or sugar is used as the source of carbohydrates. Carbohydrate and salt control the extent of fermentation as well as the keeping quality.

An alternative method is the acid pickling and storage at low temperature. This involves keeping the low-salt fermented fish with added vinegar at low temperatures, practised in the Scandinavian countries.

Fermentation without salt is not a common practice. However, alkali fermentation and fermentation by propagation of mould are practiced as some local specialties. In the alkali fermentation, often half spoiled fish is fermented using leafy plant ash as the sources of alkali. Processing of bonito by propagation of mould on the dried fish is another example of no-salt fermentation.

Another classification of fermented products is based on the technique employed in the process. Two major groups are (i) products primarily involving hydrolysis by enzymes and (ii) the product preserved by microbial fermentation. It is true that the bacteria may rapidly die off in the fermentation media containing salt; however, those capable of surviving in such media are also capable of supplying enzymatic fermentation process. It is also known that the microorganisms capable of growing in the medium contribute to the development of the characteristic aroma and flavour.

2.6.1. 1. Fish Pickle

Pickling of seafood is a recent concept and different styles of shrimp/fish pickle have been prepared to satisfy a wide range of consumers spread all over the country. Fish is rich in essential aminoacids such as lysine, methionine, throenine and isoleucine. Fish fat is highly unsaturated which contains eicoso pentaenoic acid and docosa hexaenoic acid. This is essential to control blood cholesterol level so as to prevent heart attack. Fish is also rich in minerals like calcium, potassium, sodium, magnesium and iodine. To increase the utilization of underutilized low value fishes, which are rich in above nutrients, for human consumption value added fishery products can be prepared. Fishes belongs to family leiognathides, sciaenides, nemipterus, mullets and lutjanus of marine origin, catla, rohu, catfish, common carp and tilapia of freshwater origin can be effectively utilized. Apart from fin fishes, shellfishes such as shrimps, crabs, cuttlefish, squid, chank, oysters and lobster meat can also be used. The product has a shelf life of 10-12 months at room temperature and can be used as condiments.

Fish pickle is a popular fermented fish product in india. A variety of methods are available for their preparation based on the local preference and taste. India exports fish pickles to Middle East and other parts of world to cater to the needs of the people of Indian origin. Pickle are exported in sealed glass or plastic bottles. In 1996-97, india exported pickles valued Rs.1.8 million and this market is steadily expanding. Low value fish like thread fin bream, sciaenids, mackerel, perches,etc. Are ideal raw materials for the preparation of fish pickle.

Fish is filleted and cut into small pieces. Small sized fish like anchoveilla can be used whole after washing and cleaning. The fish is mixed with salt in the ratio 1:1 (w/w) and kept for 2-3 h. The fish is then fried with minimum quantity of refined oil. Vegetable oils like refined peanut oil,olive oil, sunflower oil, palm oil, cotton seed oil, *etc.* can be used. Garlic, ginger and green chillies are gently fried in the oil. Turmeric and chilly powder are then added followed by fried fish. The contents are mixed well and sufficient quantity of boiled cooled water is added to just cover the materials. The contents are cooled and vinegar and salt are added. They are then packed in airtight glass bottles. Acid resistant pilfer proof are used to seal the bottles. Aging of pickle is very important. Aging for 2-3 months add new flavour and imparts the traditional taste to fish pickles. The shelf life of this product is 1 year under tropical ambient temperature condition.

2.6.2. Fish Sauces (Liquid Fermented Products)

Unlike many other techniques employed in fish preservation, no standard procedures have been laid down for fermented fish products. The type of raw materials, processing schedule, additives employed *etc.* will vary depending on the local tastes and preferences. The major concern in the marketability of fish sauce is the branded aroma developed by blending sauces made out of different fish.

2.6.2.1. Processing

In the production of fish sauce, fish and salt mixture is allowed to stand for extended periods until the whole flesh is converted into liquid containing amino acids and other protein breakdown products that are water-soluble. Fish of all types and even marginally spoiled fish can be used in this process. Methods of processing some of the traditional products are described below.

2.6.2.2. Nuoc-mam

The method of processing fish sauce varies with different countries. Nuoc-mam is a traditional product of Vietnam and is the favourite in Southeast Asia. Nuoc-mam is predominantly produced out of anchovies (*Stolephorus* spp.) as it is considered to yield one of the best sauces made from fish. An outline of the important steps involved is as follows:

(i) The whole fish is washed and mixed with salt in the ratio varying from 1 : 5 to 1 : 1 arranged in layers in tubs or vats made of wood or cement. Weight is placed on the top to keep the fish below the brine formed in the container. After 3-4 days a liquid containing blood and salt called 'blood picle' which oozes out is removed and kept aside. The fish is then mixed well and the blood pickle is poured over the fish to form a 10 cm layer over it and the container is kept covered.

(ii) The fish is allowed to remain in this condition for five to 18 months. The clear liquid floating on the surface is the nuoc-mam which is siphoned out. The liquid sauce is filtered and packed in bottles or earthen containers and kept exposed to sunlight until it is ripe.

A second harvest is usually possible by adding salt and water freshly to the system.

However, the overall quality and the protein content in this product will be poor.

A first grade nouc-mam should contain 20-23 g nitrogen per 1000 ml and 50 per cent of this should be formol titratable and not more than 15-20 per cent titratable as ammonia. The salt content should be 20-25 g/100 g and the pH below 6.0.

2.6.2.3. Patis

Patis is a fish sauce produced in the Philippines. Its production process is almost the same as that of nuoc-mam.

Small fish are salted whole; however the large ones are gutted and cleaned. They are then chopped to pieces of convenient size. Varying ratios of salt to fish are employed depending on fish quality. Fermentation is carried out in drums or cement tanks. Budu of Malaysia, Nam-pla of Thailand *etc.* are some other important traditional fish sauces.

2.6.2.4. Factors Affecting Processing and Quality of Fish Sauce

i) Salt

Though the growth of putrefying bacteria is prevented by the generally high salt content and the low pH of the medium, halophilic and halo-tolerant bacteria can grow well in a fermenting medium.

Salt content above 20 per cent is required to prevent the excess conversion of organic nitrogenous compounds to ammonia. Such sauces will contain only 10 per cent of the total nitrogen as ammoniacal nitrogen. When the salt content is less, this may go to very high levels. In the high-salt fermentation process, proteolysis is mainly carried out by the intestinal enzymes of fish and involvement of microorganisms or enzymes of microbial origin is minimal.

ii) Lipid Content

Fermentation is an anaerobic process. Therefore, oxidation of lipids and the consequent development of oxidised flavours can be considered not very significant. However, marginally spoiled fish which may contain some oxidised fat are often used to process fish sauce. Pelagic fish of high fat content are also very popular raw material of fish sauce. Consequently, presence of some oxidised fat in the sauce is quite natural. The oxidised fat may affect the coloured as well as aroma of the sauce.

iii) Colour

Colour of the sauce varies with the fish used in its production. Most of the well-known sauces like nuoc-mam, nam-pla and budu are principally processed out of anchoviella. The colour of sauce produced from fresh anchoviella varies from lemon yellow to light brown. Those produced from oil sardine (*Sardinella longiceps*) or mackerel (*Rastrelliger Kanagurta*) are deep brown to dark red in colour.

iv) Flavour

Perhaps the most important quality criterion of a fish sauce is its typical aroma. Several volatile and non-volatile compounds contribute to the flavour of sauce. Volatile compounds include ammonia, tri-and di-methly amines, low molecular weight aliphatic carboxylic acids and their esters. Non-volatile compounds are the protein breakdown products like peptides, peptones and amino acids, higher fatty acids and their esters, sugars and their derivatives, products or ATP breakdown *etc.*

Sauces from different fishes processed by different methods show variations in their aroma. The characteristic aroma of a branded product is often developed by blending sauces from different sources.

v) Accelerating the Rate of Fermentation

Fish sauce production is a time consuming process and often takes over a year for the process to complete. Many attempts have been made to speed up the process. A method employed is to carry out the process at accelerated temperatures upto 45°C. The fermentation process gets accelerated, however, the resultant sauce becomes unacceptable on account of the absence of the characteristic flavour of the classic sauces.

Another method of acceleration of the fermentation process tried and found partially successful is to carry out the fermentation in the traditional style for some period say 50 days, and then to continue it under accelerated conditions by artificially heating and aerating the fish/salt mixture in a concrete tank. This process will be complete in two months in place of 12 months for the traditional process.

vi) Nutritional Quality

Though it contains the protein breakdown products like amino acids, peptones *etc.*, fish sauce is not considered a nutritionally important product because its use is limited due to the high salt content and the hazards associated with the intake of high amounts of salt. Fish sauce is principally used as a condiment to flavour and salt many vegetables and animal product dishes.

2.6.3. Fermented Fish

Different from fish sauce, fermented fish is a product in which the identity of the fish is maintained to a large extent. Fermented herring and anchovy are very popular in the west; products locally called Makassar (Indonesia), buro (Philippines) or pekasam (Malaysia) are the important products of the East. Colombo-cured fish is a well-known product of India coming under this group.

2.6.3.1. Methods of Processing

i. Pickled Herring

Pickled herring (*Clupea harengus*) is dry salted with 15-36 per cent salt in barrels. The self-brine formed will cover the fish. After a while, if necessary, the barrels will be topped up with fish and brine of the same day's curing. Some sugar is also added to the salt as an aid in fermentation. Spices like pepper, mace, coriander, hops, cinnamon, ginger and cardamom are often added to improve the flavour. Benzoic acid also is occasionally added as a preservative. Fish pickled in this way can be kept for more than year at the European ambient temperature. The fish flesh becomes only moderately softened, the ripening process taking several months.

ii. Cured Anchovies

This is another delicacy popular in the Mediterranean region and is made out of *Engraulis encrasicolus*. Anchovies with very high fat content and weighing 35-40g each are considered the best for curing. The fish is nobbed, the guts being pulled out along with the head, and is salted in layers using 5-6 kg salt per each

10 kg fish. The fish is topped with a layer of salt and the fish is maintained under pressure with a weighted disc kept on the top. The fish will sink in the self-brine formed and additional fish and salt will be added a few days later. The fish will be kept in brine under pressure for at least six or seven months. During this period, water and fat pressed out of the fish form a layer of brine covered with fat. The liquid, which overflows, is collected and later used to spray the fish during cure.

Curing is carried out in sterile containers using sterile salt, and therefore, no microorganisms are involved in the process.

iii. Colombo Curing

Colombo curing is a pickling process practised in the South Canara and Malabar regions along the West Coast of India. Colombo cured products used to be processed mostly out of mackerel (*Rastrelliger kanagurta*). This fish is first gutted and washed free of dirt and slime followed by rubbing with salt (fish to salt 3:1) and kept in cement tanks. Pieces of *Garcinia cambogea* fruit are mixed with fish/ salt mixture at above 8 kg of fruit per tonne of fish. *Garcinia Cambogea* fruit is very acidic and hence it imparts acidity to the pickle and makes the product sour. It is the smoked-dried pulp of the fruit that is used for pickling. The fish remains in the tank for 3-4 months.

In another method, a smoked-dried piece of the *Garcinia cambogea* fruit is placed inside the belly flap of each fish. The fish with the fruit and salt are arranged in layers in large wooden vats. The vats are filled upto the top and kept tightly closed upto three months after which the product is marketed.

2.6.4. Fish Paste Products

Fish paste is another product involving fermentation and processed in many Asian countries. The processes employed for making all fermented fish pastes are essentially similar. Fish is mixed with salt and pounded to form a paste. The paste is sun dried for varying periods before it is sealed in containers from which air is excluded for maturing. In some cases, combination of the fish and salting precedes a period of sun drying. Moisture content of the final product is in the range 30-40 per cent.

Boiled Fish Paste

The fish is eviscerated, washed scale free, the fillets are obtained from the fish, excluding the larger bones. The minced meat can also be used for the purpose. The meat thus obtained is soaked in ice cold water for 24 hours. After soaking the meat is pressed to remove water and grinded along with the seasoning agents. The materials are common –salt (NaCl), mono-sodium-phosphates. All these ingredients are mixed thoroughly and further processed. They are pressed in the wooden press to give different shape to the product. The meat is cooked in stream for 80-90 minutes time. The streaming time depends upon the thickness of the product. After streaming the product is soaked in water for 10 minutes. The soaking gives a brilliant white to the product. The product is packed in polythene sheets. The shelf life of the product at room temperature is 4-5 days but when frozen can be stored for a longer period. This is very popular in south east Asian countries especially in china, where 90 per cent of the diversified product is of this type.

Typical among the paste fishery products are bagoong of the Philippines, ngapi of Myanmar, various mams of Vietnam, belacan of Malaysia and trassi of Indonesia. A typical process for making of fermented paste product from small shrimp practiced in Malaysia is as follows.

Prawn Paste

The best quality prawns are cleaned in portable water after peeling and beheading. The peeled material is homogenized with equal quantity of water; the ph is adjusted to 5.5 by using 20 per cent HCl. The material is treated with 0.05 per cent papane enzyme for hydrolysis of the product. The reaction takes place at 55 degree temperature and complete in 2 hours time. The partially hydrolyzed mass is concentrated in a steam jacketed vessel. Ingredients like starch (6 per cent), sugar (15 per cent), common salt (1 per cent), citric acid (0.2 per cent), MSG (0.2 per cent) and agar agar (0.5 per cent) are added and further concentrated until the desired consistency of semisolid paste is obtained.

The raw material used is usually Acetes shrimp together with small proportion of mysid shrimp. The shrimp is mixed well with salt in bamboo baskets or wooden tups. The proportion of salt to fish is generally 4-5 kg salt to 100 kg wet shrimp. The mixture is then spread out in thin layers on mats and allowed to dry in the sun. Drying is continued until around 50 per cent of the water is evaporated. During drying occasionally it is turned with a shovel to quicken the drying process. Drying may extend upto 4-8 hours. The semi-dried shrimp is minced in a mechanical mincer. The resultant paste is filled in wooden tubs/boxes under this anaerobic condition. It is then taken out of the box and spread on a mat and dried in the sun for 4-5 hours. The paste is minced again and filled in the wooden it is again minced and packed in blocks wrapped in cellophane or laminated paper. If kept under refrigeration belacan has a shelf life or several months.

The yield of the paste is 40-50 per cent of the raw shrimp weight. A good quantity belacan will have pH 7.6-7.8, moisture 20-40 per cent, ash (including salt) 20-24 per cent, salt 13-18 per cent and protein 30-40 per cent.

2.6.4.1. Principles and Methods of Preparation of Various Fish Paste Products

i. Fish Sausage

This is prepared from finely ground fish flesh of a single species or mixed, homogenized with starch, sugar, fat, spices and preservatives, and then filled in cylindrical synthetic or natural casing and pasteurized. It is a very popular product in Japan, where more than two lakh tones of this product are produces and consumed annually. Meat separated from any good quality fresh fish can be used. A good recipe is as follows: fresh picked meat : 650 gm, vanaspati: 90gms, starch (maida or maize): 90gms, sugar: 20gms, salt: 20gms, pepper powder: 2.5 gms, garlic: 1gm, chilly: 1 gm, onion : 15 gms, monosodium glutamate: 2 gm, potassium sorbate:1 gm, sodium ascorbate: 0.5 gm and crushed ice 105 gm. The ingredients are finely ground together, preferably at 18°C, stuffed into casing and ends sealed. They are then washed and pasteurised at 85 to 90°C for 45 to 50 minutes and cooled quickly in cold running water. This product keeps well for about two months at 5 to 15°C and two weeks at 28 to 30°C, depending upon the preservatives employed. A similar product containing small pieces of meat of good quality fish is generally termed fish ham. The product can be eaten as such or sliced and fried in oil.

ii. Fish Cake

Fish cake are produced in two forms. In the first method, minced fish containing salt is moulded in the shape of a cake and stored under frozen condition and it does not contain any other additives. In the second method, minced fish is mixed with starch and other ingredients and then moulded in the form of cake, the actual fish content ranges between 40-45 per cent.

iii. Surimi

Surimi refers to wet concentration of myofibrillar protein.It is a Japanese term for mechanically deboned fish mince from white fleshed fish that has been washed, refined and mixed with croprotectants for better frozen shelf life. Production of surimi is an ancient process and first invented by the Japanese more than eight centuries ago. The USA is the leading producer of surimi followed by Japan,Thailand, Korea, and South America. Surimi is an intermediate product used for the preparation of many value added products with simulated texture, flavour and appearance.

a) Conventional Method

The conventional method of surimi processing usually uses Alaskan pollack as raw material. The fish is beheaded, gutted and cleaned, and put through a belt drum-type meat separator to separate the flesh from the bone and skin. The flesh is then repeatedly washed with chilled water of 5 -10 C until odorless and colorless. It is next transferred to a strainer where residual black skin, bone and scale are removed. At this point, the flesh should be white, odorless and residue-free.

Washing plays an important role in surimi manufacturing; it not only removes undesirable matter such as blood, soluble protein and odorous substances, but also more importantly increases the concentration of actomyosin which governs the gel-forming characteristic of surimi. The concentration of actomyosin is a measure of the gel strength of surimi, an important property allowing for the development of high-quality surimi-based fishery products.

Long-term stability during frozen storage is another important characteristic of the product. For this purpose, the dehydrated material is usually mixed using a silent cutter with cryoprotectants (sugar, sorbitol and polyphosphates) at levels of 4.0, 4.0 and 0.2 per cent respectively, to prevent muscle protein from denaturation during frozen storage. Under constant low- temperature storage (below -20 C), frozen

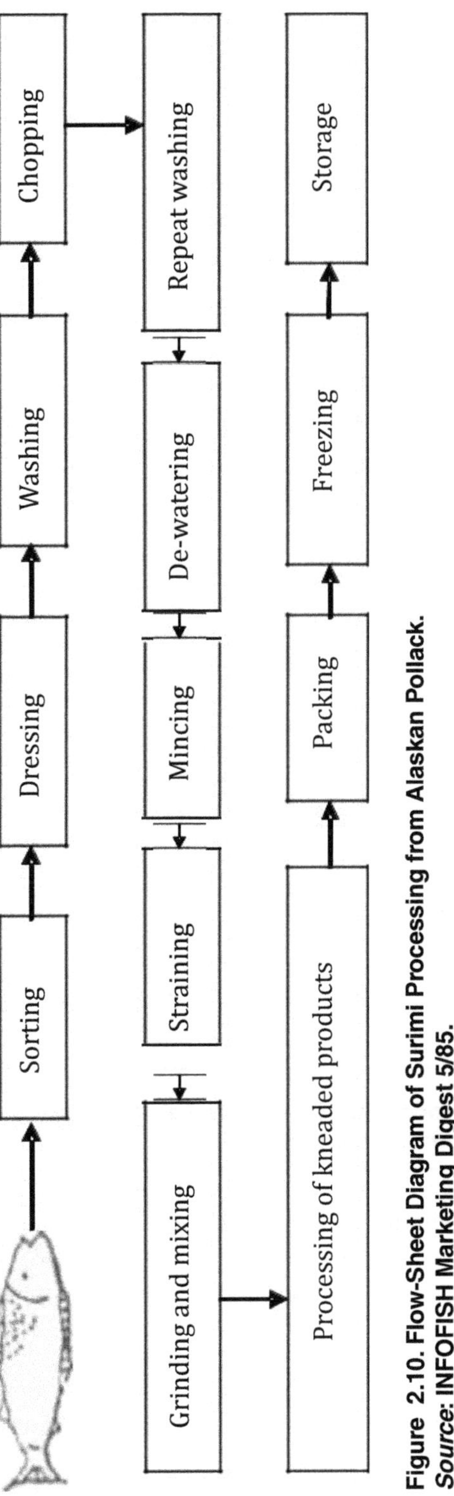

Figure 2.10. Flow-Sheet Diagram of Surimi Processing from Alaskan Pollack.
***Source*: INFOFISH Marketing Digest 5/85.**

surimi can usually be stored up to one year without significant changes in gelling properties. Nevertheless, the quality of frozen surimi is very much dependent on the freshness of raw material. Prime quality raw material is, therefore, highly desirable for surimi manufacturing.

b) *Surimi from Small Pelagic Species*

Production of frozen surimi from small pelagic species is somewhat cumbersome owing to:

a) a higher fat content;

b) instability of muscle proteins;

c) large amount of sarcoplasmic protein; and

d) a higher proportion of dark muscle in comparison with whrte muscle.

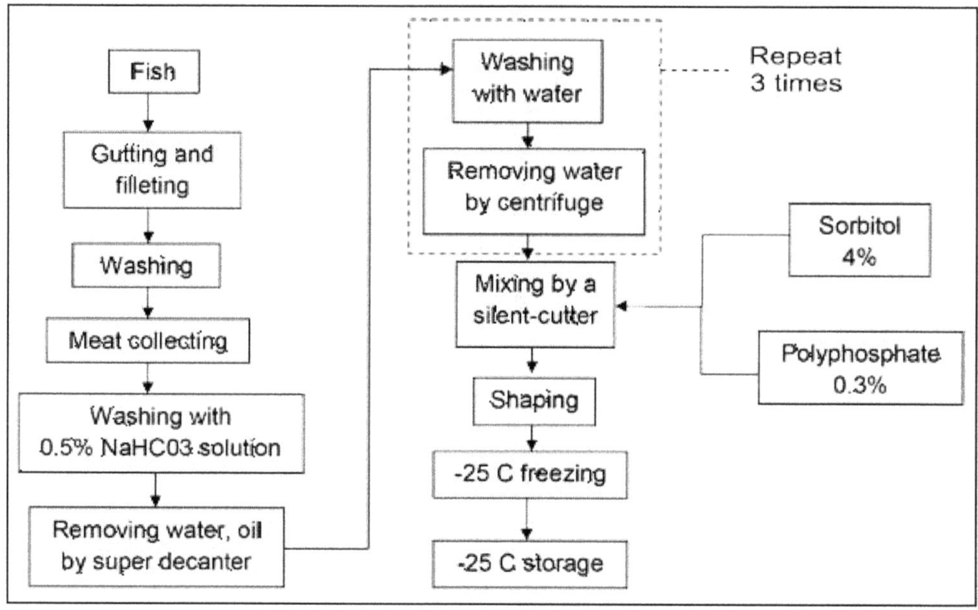

Figure 2.11. Surimi Processing Small from Pelagic Fish.
***Source*: INFOFISH International 5/89.**

A notable achievement is the development of a new processing method to these problems for small pelagic species, especially mackerel and sardine. The method involves the introduction of a meat separator into the processing line while the high fat content can be substantially reduced by washing the fish flesh with sodium bicarbonate solution followed by the use of a superdecanter to separate fish oil and water from the meat. A rotary freezer or pressure shower has been developed to mechanically remove the dark muscle, skin and subcutaneous fat and to collect ordinary muscle.

The washing process in surimi production is very effective in removing fat. About 80 per cent of the fat of the whole meat including dark muscle, and 60 per cent of the fat of the ordinary muscle is removed by this process. Washing with sodium bicarbonate solution is very effective in removing fat, followed by the use of a super decanter for separating the oil and water from the meat. Another useful method of removing fat is to pulverize the minced meat into minute particles followed by washing.

Red meat of pelagic species usually has low gel-forming ability and may not be suitable for fish jelly production. This is due partly to the low pH (below 6) of the meat and the high fat content. Adding the pH to 6.8 - 7.0 using sodium bicarbonate during the leaching process could improve gel-forming ability to some extent.

c) Surimi from Shrimp Trawl-By-Catch

Raw materials commonly used include threadfin snapper (*Nemipterus* sp.), big eye snapper (*Priacanthus* sp.), barracuda (*Sphypaena* sp.), and croaker (*Pennahia* sp.). The raw materials should be as fresh as possible, since it is impossible to obtain good quality frozen surimi from frozen fish and fish of poor quality.

Treatment of Raw Material

Since trawl by-catch usually comprises several species, it is desirable to sort the raw material according to species and sizes. The fish has also to be kept in chilled condition throughout the processing operation to avoid deterioration and improve gel-forming ability. The fish are then beheaded and the viscera are completely removed as residual enzymes can decrease the gel-forming ability of frozen surimi during frozen storage. After heading and gutting, the fish must be washed with chilled water to remove surface contamination, loose scales, blood, pieces of gut and so on. The fish should then be kept in ice prior to further processing.

Meat-Bone Separation

The large quantity and small size of raw material makes it too laborious and costly to fillet the fish manually. Therefore, the use of a meat-bone separator is more practical. There are two types of meat-bone separator available in the market — the stamp-type and roller-type. These meat-bone separators work on the principle of compressing, the fish against a steel plate or drum with mesh size of 3.5 - 4.0 mm. The processing should be kept at low temperature to minimize the deleterious effect of frictional heat on the product. The yield and quality of minced meat can be controlled by adjustment of the stamp or belt pressure. Usually the drum-type is more compact and popular.

Leaching Method

The leaching process is important to remove dark red meat, fat, blood and remnants of kidney tissue as well as off-odour, especially if the raw material used is already stale. The leaching process could be done either using simple (pails and nylon mesh) or more sophisticated equipment (rotary sieves, screw presses, *etc.*) epending on the scale of operation. The leaching process involves washing the minced meat 2-3 times with iced water (1: 5 v/v) containing 0.2-0.3 per cent salt to facilitate removal of water from the minced meat. After thorough stirring, the minced meat is allowed to settle, and the water is decanted and then drained through a nylon mesh or a rotary sieve. The minced meat is then pressed to remove excess water, until the final moisture content is about 80-82 per cent. The screw press is suitable for continuous processing of large quantities of meat whereas batch processing could be done using a centrifuge or hydraulic press.

Straining

Straining plays an important role in removing the remaining scales, connective tissues, membranes and small bones from the minced meat. The minced meat is forced through a strainer which filters the leached meat through a fine mesh of 1.2 - 3.2 mm. In order to minimize the heat generated by friction, some strainers are equipped with ice jackets to absorb heat.

Mixing

Sugar and polyphosphate are added into the minced meat as cryoprotective and water binding agents respectively and mixed using a mixer, grinder or silent cutter. The stabilised minced meat is then packed in 10-kg blocks and quick-frozen at -30°C.

d) Processing of Fish Jelly Products

In addition to imitation crab legs and shrimp, the popularity of fish jelly products is also increasing especially in countries with abundant supply of small pelagic species and trawl by-catch which are considered to be less suitable for surimi manufacturing particularly due to high content of red meat and fat. However, these raw materials have high gelling capacity and are very suitable for minced fish production as an intermediate of fish jelly products. In Southeast Asia, fish jelly products are prepared from ground fish mixed with salt to yield a smooth sticky paste. Various ingredients are then added to enhance the taste and flavour, followed by shaping and cooking. The fish jelly products industry in Southeast Asia is developing, rapidly but is still run mainly as a small household industry.

Fish jelly products are usually made of minced fish or frozen surimi. The meat is ground with salt to enhance the formation of a "complex network" and extract the salt soluble protein. Incorporation of salt also adds to the taste of the final product. Formation of a complex network is very important to give springiness after cooking. The amount of salt added is about 3-5 per cent of the weight of the meat, depending on consumer preferences. After salting, other ingredients and water are added to improve taste and texture. When the final shape of the product is formed the setting process begins. Under ambient temperature, fish paste begins to set rather quickly resulting in a firm translucent product which maintains its shape.

Therefore, it is imperative to keep the fish paste chilled throughout the processing. Traditionally, fish balls and related products are set by soaking in water prior to cooking, since the product tends to change its shape if left in air. Cooking is usually done by heating the product in the water, preferably at 90°C. The temperature at the center of the product should reach at least 80°C during the heating process. The time of heating varies according to the size of the product and should ensure complete destruction of contaminating bacteria.

Boiling of the product is not recommended, since it tends to give a rougher surface to the product.

Chemical Composition of Surimi Analogues

The chemical composition of selected surimi analogues is given in the following Table 2.6.

Table 2.6: Chemical Composition of Surimi Analogues

	Alaskan Pollack	*Surimi Pollack*	*Alaskan King Crab*	*Surimi Crab Analogue*
Protein	14.45	15.18	15.37	12.02
Calcium (mg/100g)	4.90	8.70	73.10	13.40
Potassium (mg/100g)	311.00	78.00	204.00	90.00
Niacin (mg/100g)	1.36	022	1.1	0.18
Cholesterol (mg/100g)	47.50	29.70	41.7	19.70

Source: National Food Processor Association, USA.

Quality of Surimi

The quality of surimi is determined by its gel strength and colour which are influenced by the raw material used, freshness, processing method, moisture content, control freezing and storage temperature as well as handling and distribution methods. Although surimi with lower gel strength could be used for products such as fish cakes where gel strength may not be so important, high gel strength surimi will fetch higher price and has wider acceptability under market conditions.

The following are common methods used to assess the quality of frozen surimi:

a) Sample Preparation

An aliquot sample of frozen surimi was drawn and made into paste by grinding With 2.5 per cent salt and 30 per cent chilled water for 30 minutes. The paste is then filled into a 25-35 mm diameter pvc casing and allowed to set in water bath at 40°C for 20 minutes, followed by heating at 90°C for 20 minutes. Gel strength measurement is usually performed using samples of 25-35 mm length, whereas slices of 4-5 mm are used for folding and teeth-cutting tests.

b) Gel-Strength Measurement

Gel-strength properties reflect the springiness or elasticity of the product. Together with the results of organoleptic tests it gives the quality grade of the sample.Among widely used methods of gel-strength measurement is the penetration (puncture) method using Fudoh rheometer, and elasticity measurement using Shimizu tensiometer.

The Fudoh rheometer unit contains a plunger with a sphere of 5-10 mm diameter at the tip of a 10 cm long rod. Gel-strength measurement is carried out by pressing the plunger onto the surface of the sample it until penetrates through the surface. The gel-strength is expressed in g-cm, as the product of the force required to break the sample (L, in gram) and the depth before penetration (h, in cm).

c) Whiteness Determination

This is another important quality criterion of surimi, especially for products like fish ball and fish cakes. Whiteness is usually determined using a whiteness meter, and expressed as per cent Lovibond after comparing with standard whiteness of 93 per cent Lovibond pure whiteness as shown in the following figure:

The Table 2.7 gives a comparison of the whiteness of different surimi products.

Table 2.7: Comparison of Whiteness of various Surimi Products

Species	*Whiteness (Per cent Lovibond)*	
	Minced Meat	*Frozen Surimi*
Lizard fish	41-45	58-63
Threadfin bream	32-35	53-55
Big-eye snapper	36-39	3-55 1
Glass fish	41-45	53-56
Barracuda	35-38	55-58
Croaker	30-31	53-54
Hairtail	32-35	44-47
Threadfin bream surimi from Thailand	-	50-54
Commercial fish ball from Singapore	-	46-50

Source: SEAFDEC.-937.

d) Organoleptic Tests

Although organoleptic tests are subject to the perception of the individual panelist,they provide a convenient and quick assessment of the springiness of the product.The following are organoleptic tests commonly used for measuring gel-strength of surimi:

Folding Test: This is carried out using five slices of 4-5 mm slices of sample. The slices are folded in half, and further folded into quarter. Observations and gradings are made as follows:

Condition of Test Sample when Folded	***Grade***	***Remark***
No breakage in any of five samples when folded in quarters	AA	Very good
Slight tear in any one of five samples when folded in quarters	A	Good
Slight tear in any of five samples when folded in half	B	Fair
Breakage (but 2 pieces still connected) when folded in half	C	Poor
Breaks completely into 2 pieces when folded in half	D	Very poor

Teeth Cutting Test: is carried out by biting the 4-5 mm slice samples between the upper and lower incisors. The scoring system is carried out using a hedonic scale as follows:

Score	***Description***	***Score***	***Description***
10	Extremely strong springiness	5	Acceptable, slight springiness
9	Very strong springiness	4	Weak springiness
a	Strong springiness	3	Quite weak springiness
7	Quite strong springiness	2	Very weak springiness
6	Acceptable springiness	1	Mushy texture, no springiness

Kamaboko

Surimi commercially prepared from Alaska Pollock is the important raw material from which kamaboko is made. Flat fish, lizard fish, sardine, tuna,sword fish and others are also now used for preparation of kamaboko. The surimi for the preparation of kamaboko contains a number of additives. On the contrary, if the surimi does not contain much of additives it is again converted to a fine paste by adding following ingredients.

- ☆ Common salt 3 per cent
- ☆ Monosodium glutamate, preservatives
- ☆ Wheat starch, potato starch, corn starch (5-10 per cent)
- ☆ Grinding time varies from 30-40 minutes.

2.6.5. Lactic Acid Fermented Products

Fish products fermented with lactic acid are generally processed out of freshwater species of fish. Fermentation with lactic acid is quicker than that with salt alone. Such products will contain lower levels of salt in the range 6-18 per cent compared to over 20 per cent in sauces. Their processing involves mixing the fish with salt and a carbohydrate such as cooked rice.

The keeping quality as well as the extent of acid fermentation will depend on the amount of salt and carbohydrate used. Carbohydrate is used as a food source for the lactic acid producing bacteria.

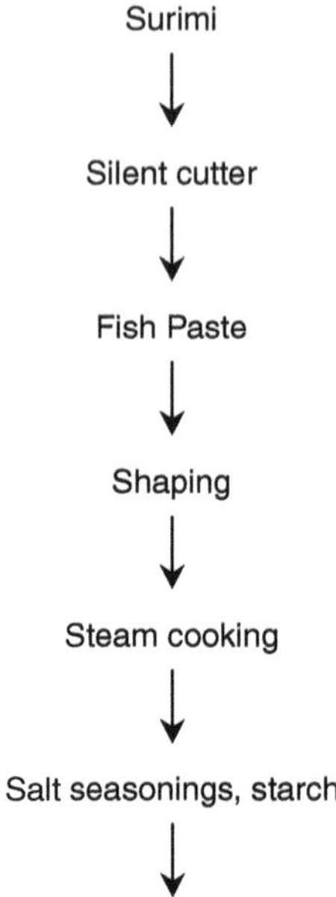

Figure 2.12. Flow Chart for Kamaboko Production.

Pla-ra is prominent lactic acid fermented product of Thailand. The process in its production involves the following stages. The process in its production involves the following stages.

The fish is scaled, eviscerated and mixed with salt in the ratio 3:1 and it then packed in jars and kept for periods extending from 15 to 90 days. After the stipulated storage period the fish is taken out of the jars, washed well and allowed to drain. It is subsequently mixed with ground roasted rice in the ratio 1:10 rice to fish and again placed in the jars for further storage for one to six months.

Consistency of *pla-ra* varies from a dark brown liquid to partly dried fish pieces. It may even be eaten as the main course.

Chapter 3

Extrusion Products

3.1. Extrusion

Extrusion is a continpous process by which moistened, expansible. Starchy, and/or proteinaceous materials are plasticized and cooked by a combination of moisture, pressure, temperature, mechanical shear. This process which combines several unit operations including mixing, cooking, kneading, shearing, shaping and forming. Extrusion cooking, which is a continuous, high- temperature, short-time process, has become popular economical process to formulate new cereal foods. Extruders are classified according to the method of operation (cold extruders or extruder-cookers) and the method of construction (single or twin-screw extruders).

3.2. Theory of Extrusion

The principles of operation are similar in all types: raw materials are fed into the extruder barrel and the screw then convey the food along it. Further down the barrel, smaller flights restrict the volume and increase the resistance to movement of the food. As a result, it fills the barrel and the spaces between the screw flights and becomes compressed. As it moves further along the barrel, the screw kneads the material into a semi-solid, plasticized mass. If the food is heated above 100°C the process is known as extrusion cooking (or hot extrusion). Here, frictional heat and any additional heating that is used cause the temperature to rise rapidly. The food is then passed to the section of the barrel having the smallest flights, where pressure and shearing is further increased. Finally, it is forced through one or more restricted openings (dies) at the discharge end of the barrel as the food emerges under pressure from the die, it expands to the final shape and cools rapidly as moisture is flashed off as steam. A variety of shapes, including rods, spheres, doughnuts, tubes, strips, squirls or shells can be formed. Typical products include a wide variety of low density, expanded snack foods and ready-to-eat (RTE) puffed cereals.

3.3. Classification of Extruded Products

Extruded products are classified into:

- First generation extruded products
- Second generation extruded products
- Third generation extruded products- protein enriched extruded products

Table 3.1: Extruded Products

Types of Product	*Example*
Cereal-based products	Expanded snack foods
	RTE and puffed breakfast cereal
	Weaning foods
	Pre-gelatinized and modified starches, dextrins
	Crisp bread and croutons
	Pasta products
	Pre-cooked composite flour
Sugar-based products	Chewing gum
	Liquorice
	Toffee, caramel, peanut brittle
	Fruit gums
Protein-based products	Texturised vegetable protein(TVP)
	Semi-moist and expanded pet foods and animal feeds and protein supplement Sausages products, frankfurters, hot dogs, Surimi Caseinates, Processed cheese

3.4. Background of Extrusion Technology

Extrude

The verb "extrude" describes a process of shaping by forcing softened or plastized material through dies or holes by pressure. A food extruder is a device that facilitates the shaping and restructuring process for food ingredients.

Extruder

Extruder is a tool used to introduce thermal and mechanical energy to food and feed ingredients, forcing the basic components of the ingredients, such as starch and protein, to undergo chemical and physical changes and form a predetermined shape.

Extrusion Cooking

Extrusion cooking is a modern HTST (High Tempreture Short Time) process is being increasing used in food and technical industries. New product are being created and well -know ones are copied and made more advantages by replacing conventional machine and equipment bya extruder cooker, which are carrying out many of the physico- chemical process. Cooking extruder are used as HTST piece of equipment for the processing of vulnerable food material.Their function is to rupture cell walls by heating and shearing, making the cell contents available for mixing and reaction with other components. The extruder function is to denature the raw material, in order to make them available for safe digestioin.In combined product industries, like the pet food industry, the summation of starch and protein-processing is executed in cooking extruder simultaneously and much emphasis is put on controlling reactions while mixing operation are executed in the extruder equipment.

The extruder still offers a small reaction volume, which can be very well controlling vapour pressure, the residence time and the viscosity of the confectionery mix in the extruder. It is the infinite flexibility of extruder design, which make them attractive as versatile production equipment in the confectionary industry. Continuous high tempreture short time extrusion cooking is not a new technology as such but an understanding of it, is of current interest. Over the last fifty year there has been an art of food extrusion and much work has done on an empirical basic studies dealing with the thermoforming of plastics. Only in recent time some of the basic thermo-mechanical transformation of food ingredients in the extruder have been understood.

3.5. Principle of Extrusion Cooking

From an engineering point of view, a cooking extruder is a combination of a pump and a heat exchanger. The raw material is fed in to the extruder through the hopper and forced forward, toward the die, by the rotation of the one or more screw. The barrel wall is generally heated by electric heater or by oil jackets, although for simple starch gelatinization adiabatic extruder are also used. The heat is provided purely by viscous dissipation.

3.6. Purpose of Extrusion

1. **Sterilization:**The heat and pressure produced in the extruder can be utilized for bacterial, mold and yeast destruction.
2. **Expansion:**The continuous pressure cooking and sudden release of pressure allows the gelatinization of starch cells, oil rupture, and shaping and texturizing the product.
3. **Dehydration:**Within certain parameters, a 50 per cent loss in moisture can be achieved through the extrusion process.This allows for dehydrating many product, which have too high a moisture level for storage.
4. **Stabilization:**The use of heat and pressure can be used to inactivate enzymes that would occur in rice bran and in various other ingredients causing rapid destruction of the nutritional properties.

3.7. Operating Characteristics

The most important operating parameters in an extruder are:

- Temperature
- Pressure
- Diameter of the die apertures
- Shear rate

The shear rate is influenced by the internal design of the barrel, its length and the speed and geometry of the screw. Most research to model extruders has been done with single- screw machines because twin-screw extruders are substantially more complex. In operation, the single-screw extruder acts as a type of pump, dragging the food through the barrel and increasing the pressure and temperature before the food is forced through the die. For optimum pumping, the food should stick to the barrel and slip freely from the screw surface. However, if food slips on the barrel it does not move through the extruder and is simply mixed.

3.8. Type of Extruder Machine

Extruders are composed of five main parts:

1. The pre-conditioning system
2. The feeding system,
3. The screw or worm,
4. The barrel,
5. The die and the cutting mechanism.

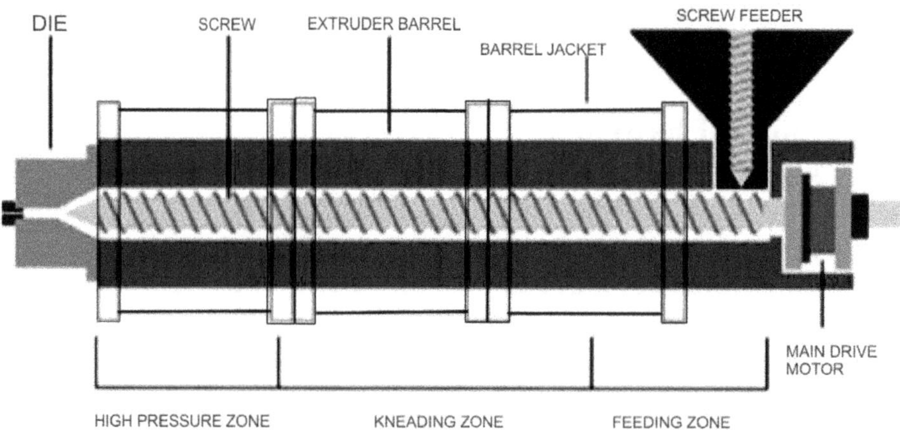

Figure 3.1. Schematic Representation of an Extruder including its Main Parts and Zones.

There are 3 major type of extruder used in the food industry:

1. Piston extruder
2. Roller type extruder
3. Screw extruder

There are three zones in single screw or twin screw extruders.

- ✰ Input zone
- ✰ Kneading zone
- ✰ Cooking zone

Input zone - where the raw materials are introduced

Kneading zone - Where the ingredients are subjected to extreme mixing, pressure and stream heat.

Cooking zone or high pressure zone- uniformly blending the dry preconditioned ingredients, injected into the extruder on a cooking-extrusion system.

1. **Piston extruder** - can consist of single piston or multiple sets of pistons that deposit a process amount of product on to conveyers or trays. Piston extruders are primarily used for forming product shape and are used in confectionary as well as bakery production.
2. **Roller type extruder** - are used to form the shape of a product. A roller consist of two counter rotating roller that either turn at similar or differential speeds. This process is also referred as calendaring in that dough industry. The roller surfaces can be smooth to create a long thin strip or can be perforated to form the dough in to shaped products.
3. **Screw extruders** - Utilize single, twin or multiple screws rotating within a metal cabinet called the barrel. The screws convey the material formed and though a small orifice called die which can take on many shapes and sizes.

Several external parameters such as screw speed and configurations, temperature of the barrel, die size and shape and the length of the barrel effect the properties of the final product.

3.8.1. Single Screw Extruder

Single screw extruder is an entry-level machine for small-scale operations. A large proportion of the expanded animal feed and modified raw materials such as soya bean and corn for feed purposes use single screw extrusion. The heat required to successfully cook the material relies upon the mechanical energy applied to the ingredients. This extruder is normally incorporated in a complete system with a hammer mill, mixer and cooler.

3.8.1.1. Types of Single Screw Extruder

- ☆ Pasta extruder
- ☆ High pressure forming extruder
- ☆ Low shear cooker extruder
- ☆ Collets extruder
- ☆ High shear extruder

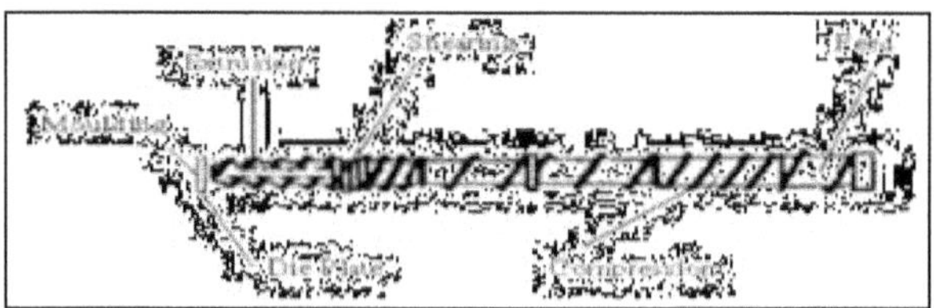

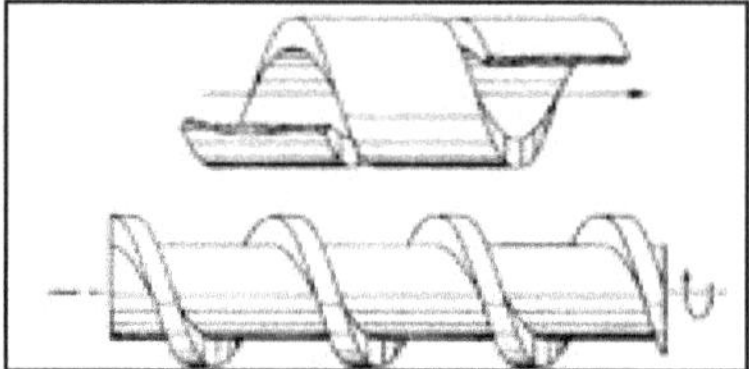

Figure 3.2. Single Screw Extruder.

3.8.2. Twin Screw Extruder

A twin-screw extruder has two screws running in parallel with intermeshing flights for increased performance. Expansion of the ingredients is still limited without the introduction of additional heat and moisture. These extruders are therefore normally supplied with either a water mixer or steam conditioner for superior gelatinisation and expansion, by raising the moisture content of the mix to 20-25 per cent prior to entering the extruder, or in the case of steam extrusion, the temperature of the material is also increased to approximately 55-65°C prior to entering the extruder barrel. This is controlled using metering systems and injected either into the preconditioned or sometimes directly into the barrel of the extruder. A higher amount of thermal energy can be supplied in this way in addition to the mechanical energy needed to cook the product.

Figure 3.3. Twin Screw Extruder.

Types of Twin Screw Extruder

- ☆ Counter rotating twine screw extruder - screws move in opposing directions
- ☆ Co-rotating twin screw extruder - screws move in same directions

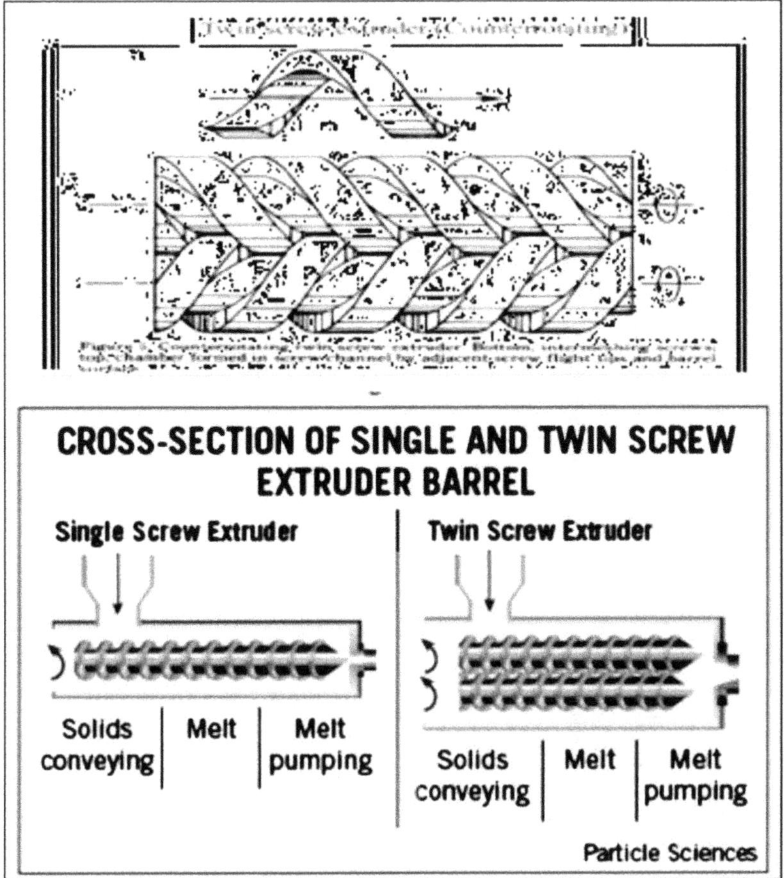

Figure 3.4. Cross Section Profile of Single and Twin Extruder.

3.9. Influence of Process on Raw Material Properties

It is important to consider ingredients properties during extrusion. Each starch source has its strong points. The fine main cereal grain starch options include rice, (long, medium or short grain) wheat (soft, hard) corn (white or yellow), barleand soats. The main objective of extrusion of starch are apart from shaping,to break down the starch granules and to gelatinize their contents. As a result of the moisture added during the process, the starch particle swell and owing to thermal influence, loss their rigidity, part of the amylose diffuses out. The high shear in the extruder rupture the particle so that all the amylose and amylopection becomes available to form an amorphous mass. Although some degredation of the large molecules can occur during extrusion, no monosaccharides are believe to be formed because of the short residence time at the high temperature.

Table 3.2: Difference between the Single Crew and Twin Screw Extruder

	Single Screw Extruder	Twin Screw Extruder
Control of product flow	Friction between the screw and the barrel is used to convey the products	The dual screw action convey the products through the barrel
Slippage and surging	The product does not undergo proper cooking or processing	Slippage occurs when high pressure in the barrel causes the product to slip between the screws and the board wall. The surging occurs when the product is held back from proper flow. Pressure builds up, and finally, the product bursts uncontrollably through the die. This can result in undercooked and misshapen products.
All the processing parameters	Can be adjust the processing factors, but to a much lesser degree	Better control of processing factors temperature, pressure, screw speed, moisture content and flow rate can be varied idependently.
Mixing	Proper mixing compared to twin screw extruder	Offer a wide range of mixing elements, including intricate mixing discs, reverse screws and compression screws that provide superior mixing
Ingredient particle size	Require uniform particle size for optimal processing	Can accept a wide range of particle sizes, assuring proper mixing and processing
Cleaning	Susceptible to dead spots where product not conveyed will adhere to the screws. This "burned on" product can break loose, block the die and plug the barrel	Self-cleaning The dual screws on a twin screw extruder are " self-wiping", that is, complimentary screw flights from one shaft remove product from the other as they turn.
Flexibility	Product produced from twine screw extruders can not be prepared using single screw extruder	Most products produced on single screw machines can be made on twin screw machines
Level of moisture and fat content	Cannot be handled foods with higher fat content and higher sugar content	Can process foods with higher fat content and higher sugar content
Scale up	Scale up the process of going from laboratory development of product to full-scale production, is mainly a single screw problem	More predictable scale-up on TSE's not a problem as long as proper data is recorded during research and development of new products. Twin screw extrusion gives you very accurate control on production parameters, which is critical in scale up.
Start-up and shut-down costs	high	Reduced to start-up and shutdown costs. Can be handled more quickly on a twin screw extruder Save time and money, especially on high volume production lines.

3.9.1. Protein

The main function of extrusion in protein processing is to denature and to texturize. As a result of the high temperature and shear forces in the extruder, the intramolecular bonds in the protein molecule are broken and the protein denatures. Subsequently, these bonds can recombine with their counter parts from other molecules thus forming an intermolecule cross-linked network in the form of strands. The shear forces in the extruder will align these strand to give the product the desire texture. For good alignment and structure, the flow conditions in the die are very important.

3.9.2. Stability of End Products

Various enzymes and microorgnisms can spoil food product if they are not inactivated or killed during processing. Lipase and lipoxygenase may, for example easily cause off flavours. Mirosinase may release the toxin isothiocyanate, and nutrient destruction can occur through various residual enzymatic activities. Enzyme inactivation therefore contributes to the storage stability and safety of the product. The influence of extrusion condition on enzyme inactivation increase with increasing extrusion temperature shear and pressure.

Table 3.3: Advantages and Limitation of Extrusion Cooking

Advantages of Extrusion Cooking	*Limitation of Extrusion Cooking*
1. Versatility	1. High capital cost
2. High productivity	2. Non availability of spare parts
3. Low cost	3. Repair and maintenance
4. Product shapes	4. Power failure
5. High product quality	5. Increased packaging cost
6. Enery efficient	6. Start-up time is high
7. Production of new food	
8. No effluents	

Table 3.4: Application of Extrusion Process in Food System

Confectionary product	Pet foods
Beverage product	Animal feed
Breakfast cereals	Sugar decrystallization
Snacks	Modified starches and starch degradation products
Texturized plant product (Texture vegetable protein)	Protein modification
Blended products	Conversion of cellulose
Pasta products	Meat analogs
Restructed muscle foods	Frozen products
Instant soup and gravy bases	Instant noodles
Conventional bakery products	Non-traditional Application of extrusion

3.10. Extrusion Application on Seafood System

Food and Agricultural Organization (FAO 1995) estimated that by the year 2010 the world's demand for fish as food for humans is 110 to 120 million tons. Ten per cent of the total catch consists of fish that are underutilized because of undesirable features like small size, dark meat, high fat content, strong flavour, high bone content, unacceptable textural properties, and/or the possible presence of toxic substances. Utilization of low-value fishes is of great importance in developing countries. Processing these fishes using high temperature or pressure, called extrusion cooking, will help modify the texture and other physical properties and make them suitable for human consumption. Co-extrusion of these fishes with cereals like rice and wheat yields nutritional products with good textural attributes and proximate composition of the extrudates.

In the recent time, extruded products are gaining wide popularity. Most of the extruded snacks are cereal and thus less protein content and are limited in some essential amino acids. The incorporation of protein rich fish mince or fish flour would increase the nutritional value of such products. Fish mince or fish flour would increase the nutrition value of such products. Fish mince or fish flour blended with cereal flour can be coextruded to obtain a nutritious snack without compromising the quality of final product. Fish which lacks good market in fresh condition, can be easily utilized for this purpose either by mincing or making fish flour. Fish are not only excellent source of high nutritional value protein but also excellent source of lipid that contain omega-3 fatty acid, especially Eicosapentaenoic acid (EPA) and Docasahexaenoic acid (DHA).

Chapter 4

Diversified Products from Fish

Fishes of low unit value having diminished demand or popularity in the fresh fihs market constitute about 50 to 70 per cent of the total marine fish landings in India. Of late, fresh fish have started to command increasing popularity and many species listed among trash fishes until a few years back are now finding place in the list of table fishes. Still many species hold out great potentialities of being converted into value-added, diversified, speciality products, in which form they can bring in better returns to the fishermen.

4.1. Breaded and Battered Products

Coating seafood poultry or vegetable products with a batter and/or breading before cooking is an established domestic as well as commercial practice. Of late, the emphasis of coated products has shifted from the home front to the restaurant and fast -food outlets. Prominent among the coated products are the seafood products. Battered and breaded fish finger or fish stick has been one of the forerunner in this field and still command a sizeable share of the battered and breaded seafood trade. Such products, when frozen in ready -to -cook form, offer a convenience of high consumer value and are called " convenience foods". Fish finger, fish portion and fish cake are the staple battered and breaded fish products. Breaded shrimp, scallop and oyster cater to luxury market.

Coating Process

The unit operations in development of coated products are portioning/forming, predusting, battering, breading, fish drying, freezing, packaging and storage.

Predusting

Predusting usually a very fine, dry,raw flour material that is sprinkled on the moist surface of the frozen or fresh food substrate before any other coating is applied. It improves adhesion of the batter, because it absorbs part of the water on the surface of the food. If the batter is applied to a surface that is too moist, it can slip, leaving some areas uncovered. Also, the use of predust tends to increase pickup. The most commonly used predust are wheat flour, gums and proteins, alone or in combination.

Batters

Batters are of two types, adhesive and tempura. The traditional adhesive batter is a fraud, basically consisting of flour and water, into which the product is dipped before it is cooked or fried. A bond between the product and the coating is formed. The proportion of batter and water is usually in the ratio of 1:2. Higher amount of water might affect the functional need of fixing the crumbs onto the batter and also will necessitate longer time to freeze the batter. The desired viscosity and pick up decide the ratio of components in the batter mix. The ingredients that constitute that the batter include starch,salt, seasonings, gums,egg and many other items. The batter also usually incorporates a leavening agent to favour expansion of the product during frying.

Breading

Breading is a serial-based coating, often of breadcrumbs.The main ingredients are almost same as for batters, and consist of flours, starch and seasonings. Breading is coarse in nature and applied to a moist or battered food product prior to cooking. Texture,mesh size,porosity and absorption are the major factors contributing to the texture of the breading. Mesh size may be coarse,medium or fine and is important in the formation of an attractive and economical coating system. Coarse particles are desirable to achieve textural targets: however, its excess use on a small surface area may cause its falling off during handling and transportation. Hence a balance among different mesh sizes is desirable. Cracker meal/traditional breading is widely used in fish products. Cracker –meal breaders, which are used to develop a cracker type, relatively hard texture, consist of unleavened flour. Home –style breadcrumbs are more porous than cracker meal and tend to absorb more oil and moisture. Japanese style crumbs, also called 'Oriental style or Panko crumb', has a characteristic flake – like elongated structure and excellent visual appeal, and provides a unique surface-texture when fried. Crumb coating are usually coloured with natural vegetable extracts to give a golden-brown appearance.These colours include paprika, annatto, turmeric or caramel.

Frying

Fat is the frying medium. Besides being the heat transfer medium, it is also a food ingredient that will influence the eating quality. Some may have specific flavour which may be carried over the product. Usually bleached and refined vegetable oil are used for frying. At the high temperature of frying, some fat may undergo changes such as polymerization. It is therefore important to use an oil of good quality for frying. According to normal manufacturing process, prefrying in oil is carried out at 180°C to 200°C for about 30 sec followed by freezing the product.

Packing

Conventional packaging materials like flexible plastic films alone are not suitable for packaging, since they provide little mechanical protection to the products. As a result, the product get damaged or broken during handling and transportation. The packaging may be a paperboard carton or a poly-lined paper bag or poly bag, which is heat sealed. The bags are designated to give support to a larger quantity of items such as fish fingers. In recent times, thermo-formed containers are commonly used for packaging coated products. These trays produced from food -grade materials are suitable for packaging breaded items both for domestic as well as export markets. Trays are made of polyvinylidene chloride, high impact polystyrene (HIP) and high-density polyethylene (HDPE) are unaffected by subzero temperatures and provide protection to the contents against desiccation and oxidation during prolonged storage.

1. Fish-Soup Powder

Any good edible fish like grouper or seer can be used. They are cooked and the edible meat separated as in the case of preparation of salad.

Recipe

Cooked meat: 1 kg, chopped onion: 500g, vanaspati: 90g, refined salt: 60g, maida: 250g, pepper powder: 15 g and monosodium glutamate: 2.5 g. The ingredients are mixed and ground thoroughly to give a homogeneous dough. It is then freeze-dried or vacuum-dried, powdered and packed in airtight containers like cans or laminated pouches, preferably under an inert atmosphere like nitrogen. The freeze-dried material packed in cans remains in good condition for more than two years. The powder is suspended in water at 10 per cent level and boiled for a minute to give a wholesome soup ready for the table. The spices can be varied to suit individual tastes.

2. Fish Flakes/Wafers

Cheaper varieties of fishes like threadfin breams, sciaenids, catfish, *etc.*, can be used for the preparation of this product. The fishes are dressed, cleaned, cooked in water for 30 minutes, cooled and edible meat alone separated.

Recipe

Cooked and picked meat: 1kg, starch (refined tapioca powder is the cheapest that can be used): 1kg, salt: 40 gm, and water: 2.5 litres. All the ingredients are homogenized into a fine slurry and poured in thin layers (1 mm) in flat aluminium trays (previously smeared with oil to prevent sticking), cooked in steam, cooled, cut into desired shape and dried. The product swells several times on frying in oil, becomes very crisp and wholesome and goes well with cocktails.

Any desired food colour and flavouring agent can be incorporated at the time of homogenizing to render the product more attractive.

3. Fish Fingers

Fish finger is a very popular product made out of the fish mince. Minced fish is mixed with water and sodium chloride (1 per cent) to get dough. Polyphosphate (0.1-0.3 per cent) is added to prevent drip loss during frozen storage. The dough is then shaped in the form of reactangular slabs and frozen. Minced fish, frozen in the form of slabs are sawed into fingers of uniform size (6x1.5x1cm). They are packed in polythene sheet, individually quick frozen and stored in cartons at -20° C to -30° C. The fingers are deep fried prior to use.

4. Fish Cutlet

Fish cutlet is another delicacy product among fish consumers. The basic raw material required for preparation of this product is cooked fish or mince (*fish kheema*). *Kheema* is the fish meal picked from whole fish by means of a meat picking machine.

Method of Preparation

Minced fish (65 per cent of the total weight or mix) is cooked in boiling water for 20 min and excess water is removed. Salt and turmeric are added to the cooked meat and mixed well. Boiled and peeled potatoes are made into a paste and mixed with cooked fish. Onion, garlic and chillies are chopped into small sizes and fried in refined oil until golden brown in color and are added to the mixture. The whole mass is heated again for 30 min. Powdered spices are added and mixed well. The mix is moulded into 40g size cutlets in oval or round form. They are then dipped in a batter of egg white and rolled over bread crumbs. The cutlets are then flash dried at 160-170°C for 5 sec and then held in frozen storage at -20°C. Shelf life of frozen cutlet is 22 weeks. They are thawed and fried in oil before us.

4.2. Imitation Products

Several value added imitation products are made from surimi(water washed mince). These include imitation of shrimp lobster tails breaded scallops, imitation breaded crab claws, sushi products, sushi sticks, imitation crab shreds, mincec sticks, filament sticks and others. These seafood analogs possess the accepted texture, flavour and appearance of the authentic products. For production of seafood analogs, the surimi blocks are chopped to create a paste. The paste is combined usually with additional amounts of cryoprotectants and other additive such as salt,soy protein, starch, egg white, alginate *etc.* to improve cohesion among the protein molecules and thereby to improve the texture and flavour of the finished product. For chopping the frozen blocks, it is ideal to use a vaccum mixer, which helps to disintegrate the mince and make the proteins available for binding the ingredients. A vaccum mixer also removes any air that could be introduced in the chopped product, which can result in uneven heating during cooking. The chopped paste is extruded as a flat sheet (approximately 1-2mm thick) molded into desired shapes and set by placing on a cooking belt where it is heated. Heating is done at 90-93°C for 30-100 sec on a stainless steel, belt, drum. Final texture is developed during thermal pasteurization, which is performed after bundling, cutting and packaging. The pasteurization step eliminates bacterial pathogens that might grow during the storage of the product. Generally surimi seafood should be cooled from 60 to 21.1°C or below within 2 h and 4.4°C or below within 4h and should be held at 4.4°C or below at all times during storage and distribution.

4.3. HACCP in Product Preparation

HACCP is a total quality management system with emphasis on safety based on a systematic approach to identification, assessment and control of hazards. It is a preventive control system in which hazard is controlled or eliminated before it occurs. It concentrates on prevention strategies on known hazards and the risks arising out of them occurring at specific points in the processing schedule.

HACCP Concept

Food must be safe to consume and conform to certain standards. If some properties are monitored by plant, but without supplementary information, the tests will provide only a poor means for controlling and operation. If the product does not conform to specifications, it may have to be reprocessed or discarded. This contingency can be avoided if certain key variables in the process are monitored and controlled. Such a system is provided by HACCP. HACCP is based on a set of seven principles. The system envisages identification of potential hazards in seafood processing at all stages upto the point of consumption. The seven principles are

- ☆ Hazard analysis- Assess the hazards associated with capture, storage, raw materials and ingredients, pre-process and process operations, and all other activities upto consumption. Prepare a flow diagram of the steps in the process. Identify and list hazards and specify control measures.
- ☆ Determination of critical control points (CCP)
- ☆ Specification of criteria – Establish target levels and critical limits that must be met to ensure that each CCP is under control.
- ☆ Establishment of procedure and monitoring system to ensure control of the CCP and their implementation.
- ☆ Corrective action when the monitor indicates any deviation from the critical limits or that the process is out of control.
- ☆ Establishment of procedures to verify that the HACCP is working correctly and effectively.
- ☆ Establishment of documents concerning all procedures and records appropriate to these principles and their application.

Hazard

A hazard is a biological, chemical or physical factor that has the potential to cause an adverse effect on human health. Biological hazards include pathogenic microorganisms, parasites, toxigenic plants, animals and products of decomposition like histamine. Pesticides, detergents, antibiotics, heavy metals, non-permitted food colours and food additives *etc.* constitute the important chemical hazards. Extraneous matter like filth, metal or glass fragments,stones *etc.* are important among physical hazards.

Hazard Analysis

This is a system using which the significance of a hazard to consumer safety can be analysed. By using this system it can be decided which hazards are of such nature that their elimination or reduction to acceptable levels is essential to produce a safe food product. Identification of hazards, their assessment and identification of control measures constitute the important functions of hazard analysis.

Critical Control Points

A critical control point (CCP) is a point or a stage in the processing operation where failure to control effectively would most likely result in the production of defective/unsafe food. In other words, it is a step which, if properly controlled, will eliminate or reduce a hazard to an acceptable level. A step/point is any stage in the production. This includes raw materials, transport to processing plants, processing and storage. CCPs require constant checking to ensure compliance with all the requirements of the product.

Control Points

These are other points in the processing operation where failure to effectively control may not necessarily result in the production of defective/unsafe food. These points require occasional checking throughout the production shift.

Determination of CCPs

The relevance of each identified hazard in each stage of the process is considered. If the hazard can be reduced, prevented or eliminated through some form of control at a particular stage, it is a CCP. There are two types of CCPs, CCP1 and CCP2. The CCP1 will ensure the control of a hazard, whereas CCP2 will minimise the hazard, but will not assure its control. Both are important and should be controlled. A judgement of risk also must be made so that a level of concern can be ascribed to it. The four levels of concern are High concern, where without control there is a life threatening risk.

- Medium concern, where there is a threat to the consumer that must be controlled.
- Low concern, where there is little threat to the consumer; still is advantageous to control it.

- ☆ No concern, where there is no threat to the consumer.
- ☆ The points which are not critical because of low risks need less control and monitoring. If a hazard can be controlled at more than one point, the most effective place for control should be decided.

Specification of Criteria for Control

The team should next identify the means by which the hazard can be controlled at each CCP. These may include level of chlorination in wash water, temperature during storage, moisture content, level of toxic materials, sensory parameters, time and temperature requirements for thermally processed foods *etc.* All these must be documented as clear unambiguous statements or included as specifications in operating manuals. The team will also have to set target levels and specified tolerances of the control measures at each CCP. Critical limits may also be derived from government regulations and guidelines, international codes of practices, experimental studies or any other recognised source.

Monitoring and Checking System

There must be a mechanism to monitor, check and measure to ensure that processing procedures at each CCP are under control. The monitoring must be able to detect any deviation from these specifications. It should be rapid enough so that corrective action can be taken in time and loss of the product is avoided or minimised. The main methods to monitor a CCP are visual observation.

- ☆ Sensory evaluation
- ☆ Physical measurement
- ☆ Chemical testing
- ☆ Microbiological analysis

Visual observation is basic, but gives rapid results. Sensory evaluation can be used to check the quality of incoming raw materials. Rapid chemical tests *e.g.* chlorine in water, are useful to monitor CCPs. Other measurements possible are time, temperature, pH and salt concentration. Microbiological test is of limited use in monitoring CCP. It can, however, be employed for testing before starting processing, and for testing finished products before release.

Corrective Action, Verification and Documentation

Corrective Action

If monitoring indicates that there are deviations from the critical limits or that the process is out of control, corrective actions must be taken immediately. The corrective actions must be based on the assessment of hazards, risk and severity, and on the final use of the product. A plan should have been prepared in advance so that there will be no delay in taking corrective action. The team should prepare this plan specifying the corrective action, identify the persons to implement them and disposition actions needed to be taken with the food that has been produced during the 'out of control' period.

Verification

Once the HACCP system has been drawn up for a product/process it must be reviewed before it is installed and regularly reviewed while it is in operation. The appropriateness of the CCPs and control criteria can be determined and the extent and effectiveness of monitoring can be verified. The team should describe in detail the methods and procedures to be used to verify the system. Some of the methods that can be used in verification are Reviewing the HACCP study and its records

- Random sampling and analysis (microbiological)
- Detailed tests at selected CCPs
- Survey of conditions during storage, distribution, sale and use of products
- Interviewing staff

Documentation

Proper records should be maintained on all actions in the HACCP system in order that origin, cause and point of occurrence of the hazard can be traced. Records should be maintained on the following aspects HACCP plan and supporting documentation:

- Monitoring CCPs
- Records of corrective actions
- Records of verification activities and modifications
- Nature, coding and disposition of the product.

References

A. S. Ninawe and K. Ratnakumar (2008). Fish Processing Technology and Product Development. Narendra Publishing House. New Delhi.

D.D. Nambudiri and K.V.Peter (2012). Advances in Harvest and Post Harvest Technology of Fish. NIPA, New Delhi.

K.K. Balachandran (2001). Post Harvest Technology of Fish and Fishery Products. Daya Publishing House. New Delhi.

Jhari Sahoo and M.K. Chatli (2015). Textbook on Meat, Poultry and Fish Technology. Astral International (P) Ltd., New Delhi.

K.P. Biswas (2014). Fish Processing and Preservation. Astral International (P) Ltd. New Delhi.

www.ingramcontent.com/pod-product-compliance
Ingram Content Group UK Ltd.
Pitfield, Milton Keynes, MK11 3LW, UK
UKHW021956270726
14060UKWH00002B/539